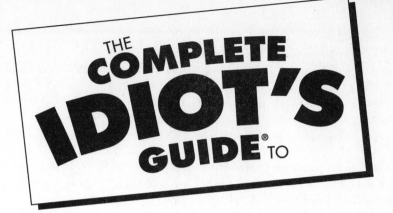

THE COMPLETE IDIOT'S GUIDE® TO

Geography

Third Edition

by Thomas E. Sherer Jr., Thom Werthman, and Joseph Gonzalez

ALPHA

A member of Penguin Group (USA) Inc.

ALPHA BOOKS

Published by the Penguin Group

Penguin Group (USA) Inc., 375 Hudson Street, New York, New York 10014, USA

Penguin Group (Canada), 90 Eglinton Avenue East, Suite 700, Toronto, Ontario M4P 2Y3, Canada (a division of Pearson Penguin Canada Inc.)

Penguin Books Ltd., 80 Strand, London WC2R 0RL, England

Penguin Ireland, 25 St. Stephen's Green, Dublin 2, Ireland (a division of Penguin Books Ltd.)

Penguin Group (Australia), 250 Camberwell Road, Camberwell, Victoria 3124, Australia (a division of Pearson Australia Group Pty. Ltd.)

Penguin Books India Pvt. Ltd., 11 Community Centre, Panchsheel Park, New Delhi—110 017, India

Penguin Group (NZ), 67 Apollo Drive, Rosedale, North Shore, Auckland 1311, New Zealand (a division of Pearson New Zealand Ltd.)

Penguin Books (South Africa) (Pty.) Ltd., 24 Sturdee Avenue, Rosebank, Johannesburg 2196, South Africa

Penguin Books Ltd., Registered Offices: 80 Strand, London WC2R 0RL, England

International Standard Book Number: 978-159257-663-0
Library of Congress Catalog Card Number: 2006920584

09 08 07   8 7 6 5 4 3 2 1

Interpretation of the printing code: The rightmost number of the first series of numbers is the year of the book's printing; the rightmost number of the second series of numbers is the number of the book's printing. For example, a printing code of 07-1 shows that the first printing occurred in 2007.

Printed in the United States of America

Most Alpha books are available at special quantity discounts for bulk purchases for sales promotions, premiums, fund-raising, or educational use. Special books, or book excerpts, can also be created to fit specific needs.

For details, write: Special Markets, Alpha Books, 375 Hudson Street, New York, NY 10014.

Publisher: *Marie Butler-Knight*
Editorial Director: *Mike Sanders*
Managing Editor: *Billy Fields*
Acquisitions Editor: *Tom Stevens*
Development Editor: *Michael Thomas*
Production Editor: *Kayla Dugger*
Copy Editor: *Michael Dietsch*

Cartoonist: *Richard King*
Cover Designer: *Bill Thomas*
Book Designer: *Trina Wurst*
Indexer: *Brad Herriman*
Layout: *Brian Massey*
Proofreaders: *Kathy Bidwell, Mary Hunt*

Contents at a Glance

Contents

Introduction

Thank you for purchasing the new, revised edition of *The Complete Idiot's Guide to Geography, Third Edition*. Please don't be put off by its title: we're all idiots when it comes to geography! Especially in a world that is changing so quickly, even professional geographers and mapmakers get stumped every so often and have to look things up in an atlas, just like everyone else! There's simply too much to geography and too much going on in the world today to be able to get a handle on it all—but that doesn't mean, of course, that you can't try.

This book is designed to give you the basics. It tells you what geography is all about and how it applies to your life. You will learn a good deal about the natural world and get a feel for the people who inhabit the many environments of our planet. You'll also learn a little about the history, culture, and politics of the world's regions. Best of all, you won't feel like an idiot the next time someone drops an exotic place name at a party or refers to a current global hot spot. If you are a serious student of human geography, you will find this text most useful as a study guide.

There is nothing more fascinating than the world we live in and the many different peoples we share it with. I hope that this book will set you on a lifelong quest for geographic knowledge. The Internet has become a terrific source of quick information on all aspects of our planet—its geography, resources, environments, peoples, history, cultures, current events, and more. To continue your geographic quest, check out such websites as geography.about.com, www.geographic.org, www.geohive.com, and the invaluable CIA factbook site, www.cia.gov. Subscribe to a geographic magazine, buy yourself a good atlas (no home should be without one!), but most of all—travel! Nothing brings geography to life as much as actually seeing places you've read about up close and in person.

The goal of this book is to give you a basic understanding of geography and an overview of the world's lands and peoples. But it is just an overview.

What You'll Learn in This Book

The Complete Idiot's Guide to Geography, Third Edition, divides your geographic journey into five parts.

Part 1, "An Overview of Geography," defines what geography is and introduces the nuts and bolts of the discipline. It lays out the basic terminology and principles you'll come across throughout the rest of the book. Part 1 is the foundation on which your geographic knowledge will be built.

In **Part 2, "A Regional Look: The Developed World,"** you begin your journey around the globe, starting with the more economically developed nations of the world. Eight regions and more than two dozen countries are included in this section. Each regional chapter surveys the terrain, environments, climates, history, peoples, and cultures of the countries that make up the region. In each chapter, you'll find the basic facts about a country or group of countries—capitals, largest cities, tallest mountains, longest rivers, ethnic breakdowns, economic activities, and current events—plus a wealth of surprising statistics and interesting background information. Don't get overwhelmed or feel obliged to memorize it all—that's what atlases and almanacs are for.

The content and format of **Part 3, "A Regional Look: Nations in Transition,"** are the same as in Part 2, but this part's focus is on three regions that don't fit neatly into the group of developed countries covered in Part 2 or the developing countries surveyed in Part 4. The regions you'll explore in this section of the book—Eastern Europe, Russia, and China—have one basic feature in common: they are all in various stages of a transition from a state-controlled communist economy to an open, free-market capitalist economy.

Part 4, "A Regional Look: The Developing World," takes you on a whirlwind tour through nine regions and dozens of countries whose economies are on the lower rungs of the economic development scale. The countries you'll explore in this section range from the very poor and troubled to some of the most exciting and vibrant places on the planet. Brace yourself for lots of color, contrast, and controversy!

The shortest part of the book, **Part 5, "A Global Overview,"** looks at problems that concern every region on Earth—environmental and economic issues that have an impact on all of us and call for global cooperation to resolve.

Extras

Throughout the five parts of this book, you'll find a variety of features designed to make your journey around the world easier and more fun. Each chapter includes several simplified maps of the region being surveyed. Although you won't want to navigate by them, the maps will give you a very good idea of what is located where. In each chapter, you'll also find sidebars and feature boxes with extra information that clarifies or expands on subjects covered in the main text. Here are the road signs you'll encounter along the way:

def•i•ni•tion

In these boxes, you'll find handy definitions of obscure or unfamiliar geographic terms.

GeoRecord

Highest, widest, deepest, longest—here are some of the superlatives of world geography.

Geographically Speaking

These boxes examine geographic, historical, or cultural topics in greater depth than is possible in the main text. Everything is fair game in these sidebars, from an explanation of plate tectonics to a closer look at India's caste system.

Terra-Trivia

Impress family and friends with these fascinating facts and bits of interesting information about people, places, and events around the world.

Eye on the Environment

The health of our planet should be a top concern of everyone who lives on it. Environmental issues are the focus of these sidebars, from the unintended consequences of Egypt's Aswan Dam to New Zealand's pioneering role in the use of Earth-friendly geothermal energy.

Acknowledgments

Without a first edition, there would be no revised edition. As the author of the new revised edition of *The Complete Idiot's Guide to Geography*, I would like to thank the authors of the previous editions, Thomas E. Sherer Jr., and Joseph Gonzalez, for laying down such a solid foundation for this book. I would also like to thank my AP Human Geography students for always keeping me on my toes, researching, answering questions, and growing in knowledge. Life is an adventure and each person we encounter is a wonderful part of the map we travel. The maps in this book were created quickly and beautifully by Nicole and Mario Antonetti, under the direction of Ortelius Design. My thanks to them for a great job on short notice. For information about Ortelius Design, please contact them at 608-835-8301 or gary@orteliusdesign.com.

Trademarks

All terms mentioned in this book that are known to be or are suspected of being trademarks or service marks have been appropriately capitalized. Alpha Books and Penguin Group (USA) Inc. cannot attest to the accuracy of this information. Use of a term in this book should not be regarded as affecting the validity of any trademark or service mark.

Part 1

An Overview of Geography

One of the most difficult challenges geographers face is defining exactly what geography means. Part 1 of this book tries to do just that. It fleshes out the meaning and scope of the term *geography* and gives you some basic background in physical, cultural, and regional geography. Along the way, you even get to refresh your map skills.

Although all this might sound like a great deal of heavy-duty ground to cover in just a few chapters, don't let that daunt you. The knowledge and skills you'll gain will be very useful as you set forth on your global explorations later in this book. With maps in hand and a newfound understanding of what geography really is, you'll be ready to take on the world.

"You know how this thing works? There a plug or something?"

1

What Is Geography?

In This Chapter

- ◆ Geography defined
- ◆ How to study geography
- ◆ Why you need to know this stuff

Geography is a difficult discipline to define, because it doesn't fit neatly into any academic box. However, to put it in its proper academic perspective, geography is about the world we live in and the people with whom we share this planet.

Geography has been called "the mother of all sciences" and "the science of place." However you describe it, geography involves the study of the physical and cultural factors that interact to make up the diversity of the earth's places and environments.

The Mother of All Sciences

Almost anything can fall into the realm of geographic study; however, the geographic discipline is basically divided into physical geography (such as geology, topography, mineralogy, and meteorology) and cultural or human geography (the accomplishments of humankind, including culture, religion, politics, economics, and architecture to name a few). Geography is a spatial discipline, which means that geographers are concerned not only with what something is but also with the way it is distributed in space.

Terra-Trivia

The Greek scientist Eratosthenes was the first to use the term geography, which literally means "to describe the earth."

Although knowing the names of places and where they are located is important to geography, it is not the sum of geographic study. That being said, it is becoming ever more necessary, as our world becomes a smaller, more interdependent place.

Why We're Geographically Illiterate

Most people in the United States got their limited dose of geography in a social studies class because most states did away with geography as a distinct course decades ago.

Time and again we find that the average U.S. citizen is geographically unaware. Although being a "geographically illiterate" nation is a good thing when you're writing a book like this one, it's bad news for U.S. business.

Virtually every other industrialized nation on Earth features geographic education as a primary component of its academic instruction, and not just an add-on chapter in some history class. In most countries, as it was in the United States just a few decades ago, geography is taught from elementary school through high school as a separate and important field of study. When graduates in those countries enter the global marketplace, they're familiar with resources, transportation systems, landforms, foreign languages, religions and customs, climate, demographics, and political systems. Most U.S. high school, and even college, graduates cannot say the same. In fact, many U.S. students don't even know their own country very well.

Thankfully, some of our education leaders are realizing that deficit and geographic study is making something of a comeback in the U.S. educational system. Prompted in part by a National Geographic Society survey in 1988 that found that one in seven young adults in the United States could not locate their own country on a map, educators are placing greater emphasis on geography and teaching it once again as a stand-alone course.

Why Study Geography?

Of course, we shouldn't be studying geography simply because everyone else is doing it. Yet, in a real sense, that is exactly why we need to be encouraging greater geographic knowledge in our population. Our world is becoming more and more unified through transportation and communication networks. We can't live in isolation, and we need to be competitive in order to attain success.

Other countries' students are entering the global workforce better prepared than our students. Those with geographic knowledge are aware of the planet's kaleidoscope of peoples, cultures, and ways of life. To a great extent, geography shapes history and current events. In our increasingly globalized and interconnected world, knowledge of geography is more important than ever to understanding the world and your place in it.

The Least You Need to Know

- Geography is a spatial discipline that encompasses "anything that can be mapped."

- There's more to geography than place names.

- The study of geography becomes ever more important in an increasingly globalized and interconnected world.

2

Physical Geography: A Look at Mother Earth

In This Chapter

- ◆ How the sun is responsible for day and night and the seasons
- ◆ How geography affects the climate
- ◆ Whether the earth is really on solid ground
- ◆ The earth's moving surface
- ◆ How plate movement and volcanoes build up the earth's surface
- ◆ How water and ice wear down the earth's surface

Physical geography is the study of the physical features and phenomena of the earth. The study of physical geography encompasses many scientific fields including hydrology (water), meteorology (atmosphere), and biology (living things). Virtually every other "ology" also has some part in the study of our planet.

It All Revolves Around the Sun: Day and Night

The source of all energy on Earth is the sun. Without it, there would be no day and night and no seasons. Life could not exist on this planet without the sun.

A flashlight aimed at a tennis ball sums up the relationship between the sun and the earth. When the light (the sun) shines on one side of the tennis ball (the earth), the other side of the ball is dark. Therefore, half the earth is always illuminated, and half the earth is always in the dark.

def•i•ni•tion

Rotation is the spinning of the earth on its axis; a complete rotation occurs once every 24 hours. Revolution describes the earth's orbit around the sun; the earth completes a full revolution once every year (365 days, 5 hours, 49 minutes, and 12 seconds to be exact)—except in leap years, which occur every four years to take care of the difference.

The earth constantly rotates, so one side isn't always in the light (or dark); this 24-hour spin causes our day and night cycle.

As the earth *rotates* (or spins), causing day and night, it also *revolves* (or orbits) around the sun. Because the earth's axis is inclined, or tilted ($23\frac{1}{2}$ degrees), sometimes the Northern Hemisphere tilts toward the sun (in June), and sometimes the Southern Hemisphere does (in December).

The hemisphere tilted toward the sun experiences summer, and the hemisphere tilted away from the sun experiences winter. The earth's tilt causes one hemisphere to experience more sunlight or less sunlight at any given time of the year, which causes summer days to be longer and winter days to be shorter. As a result, the North Pole experiences total darkness in December, while the South Pole is in total sunlight; the opposite is true in June.

Considering Climate

Although people usually use the terms weather and climate interchangeably, they mean different things. Weather refers to the here and now; it is the temperature, wind, and moisture of a specific place at a specific time. Weather is localized, it changes, and it can be unpredictable.

Climate refers to the long term, or to average weather conditions over an extended period for a large region of the earth. The world can be divided into many different climatic zones that correspond roughly to their distance from the equator. (Other factors, such as altitude and proximity to a large body of water, also play a role in determining climate.)

German climatologist Wladimir Köppen (1846–1940) divided the world into five climate zones, each with many subdivisions. (A sixth zone—mountain—was added later.)

- **Tropical, rainy:** Near the equator

- **Dry, desert:** In bands just north and south of the equator

- **Humid, temperate:** Farther to the north and south than the dry, desert areas (the Mediterranean is an important subdivision of this climate zone)

- **Humid, cold:** Mainly north of the northern temperate zone, plus the southern tip of South America

- **Polar:** Toward each pole

- **Mountain:** The upper reaches of the world's great mountain chains: the Rockies, Andes, Alps, Caucasus, Himalayas

Why Is It Hot at the Equator and Cold at the Poles?

The earth is hottest at the point where the sun's rays strike it most directly. At the equator, the sun is always close to being directly overhead. At the North and South poles, the sun is never directly overhead. During the Arctic and Antarctic winter months, the sun does not even appear. Even at the height of polar summer, the sun remains close to the horizon.

The air temperature is caused only indirectly by the sun's rays. The air is actually heated in a process called *reradiation*. As the sun's rays penetrate the earth's atmosphere and strike the earth's surface, the surface absorbs the light energy from the sun and changes it to heat energy that warms the air.

Because some surfaces are more effective than others at absorbing and radiating heat, the earth has varying air temperatures across the globe. The best absorbers of the sun's energy, and hence the warmer areas, are the vast oceans, and the dark surfaces, such as asphalt, soil, and leaves. Smooth and light-colored surfaces, such as ice and snow, are less effective in absorbing energy from the sun and cause cooler air temperatures.

Why Does the Wind Blow?

The wind blows because of differing air pressures at various locations at the same time, which is caused by the differing temperatures and moisture content of air in different places. Because water molecules are less massive than the gas molecules in the air, increased water vapor results in decreased atmospheric pressure. In addition, cold air is more compressed and contains more molecules per cubic inch, so it is heavier than warm air and yields higher atmospheric pressures.

Elevation, or how high you are in the atmosphere, also plays a part in wind formation. At sea level, the weight of the atmosphere is about 14.7 lbs on every square inch. The higher you go from sea level, as in up a mountain, the less pressure is exerted by the atmosphere.

Winds are caused when heavier, high-pressure air pushes into areas of warmer, lighter air. All things in nature seek balance, so the winds are nature's way of equalizing the pressure of the atmosphere. The greater the difference in pressure (and temperature), the faster and stronger the wind blows. Extreme differences in air pressure result in tornadoes and hurricanes.

They Call the Wind Many Things

Twisters, tornadoes, and whirlwinds all refer to isolated points of extreme low pressure (warm, moist air) surrounded by rapidly spinning columns of wind. Although these roaring, funnel-shaped vortexes are not generally more than a couple hundred yards wide, they leave terrible destruction in their path when they touch down on the ground surface. (They may attain a wind speed of 300 mph, but are more commonly in the 110 to 200 mph range.) The most tornado-prone area in the world is the central plains states of the United States, or "Tornado Alley."

Waterspouts and dust devils are miniature and often short-lived versions of tornadoes. Waterspouts are misty columns of spray over bodies of water. Dust devils are similar except that they occur over land. Hurricanes and typhoons are severe tropical storms with sustained winds of 74 mph or higher. These massive, circular- or oval-shaped storms can measure more than 300 miles across. A peculiar characteristic of these storms is their eye, an area of calm (often complete with blue sky) directly in the center of the storm. They are associated with extreme low pressure. Hurricanes are storms that occur in North or Central America; typhoons occur in the western portions of the Pacific Ocean. The intense rains and ocean storm surges associated with these storms can inflict horrific damage, such as that from Hurricane Katrina and the New Orleans flooding of 2005. "Cyclone" is an umbrella term for all the storms mentioned in this list (except for monsoons). Even low-pressure centers that do not develop into one of these types of severe storms are considered cyclones. The term "tropical cyclone" is also used more specifically to describe the Indian Ocean equivalent of a hurricane or typhoon.

Terra-Trivia

In the Northern Hemisphere, cyclonic winds spin counter-clockwise. In the Southern Hemisphere, cyclonic winds spin in a clockwise direction.

Monsoons differ from other weather phenomena in that they're not a particular type of storm. Instead, the term refers to a seasonal shifting of winds. Although generally associated with Southern Asia, monsoonal climates are also found in Northern Australia, Western Africa, and other parts of the world. Monsoons may be "wet" or "dry." The "wet summer" phase of the Indian Ocean monsoon lasts from June to September and often brings torrential rains. Typical of this phenomenon is the 37.1 inches of rainfall on Mumbai, India, in a single day on July 26, 2005.

The Earth's Moving Surface—It's Not Solid Ground!

Our earth's surface, although it appears to be solid, is really the equivalent of many giant rafts moving and bumping around on an orb of molten material while floating around in space.

Two hundred twenty-five million years ago, all the continents were part of a huge supercontinent called Pangaea (Greek for "all earth"). About 180 million years ago, Pangaea began to split up. Its parts slowly drifted away from, and in some cases toward, each other. The process, called continental drift (a term coined by Alfred Wegener), still continues. The earth's surface, both the land masses and the ocean floor, is made up of giant plates or solid chunks, which is explained in the theory of plate tectonics.

Floating and Colliding Continents

According to the theory of plate tectonics, the earth's outer shell, called the lithosphere, is not continuous, but instead is divided into irregularly shaped rigid plates that float on an underlying molten layer called the asthenosphere. The plates move relative to each other at rates of as much as several inches a year, and they meet in a number of different ways.

At divergent boundaries, adjacent plates are actually forced apart by molten matter rising up from the asthenosphere to form ocean floor ridges. The Mid-Atlantic Ridge, which splits nearly the entire Atlantic Ocean floor in half north to south, is a good example of a divergent plate boundary.

When plates collide at what are called convergent boundaries, the older, heavier plate is forced to sink below the lighter, younger one, often forming deep ocean trenches. This process is known as subduction. If the colliding plates are carrying continents that also bang up against each other, the result is new mountain formation at the edge of the lighter, younger plate. The Himalayas, for example, are the result of the collision between the Eurasian Plate and the Indo-Australian Plate.

GeoRecord

Some of the world's deepest subduction trenches are in the Pacific. The Mariana Trench in the Western Pacific is the deepest spot on Earth. There, in an abyss called Challenger Deep, the earth's surface plunges to 35,826 feet below sea level.

Tectonic plates also meet at transform boundaries. In this case, the plates don't collide head-to-head, but rather grind sideways against each other, creating lateral faults (such as California's famous San Andreas Fault) and earthquakes.

All of this plate movement generates tremendous amounts of energy, which is released in part through volcanoes and earthquakes. In fact, most of the world's volcanic and seismic (earthquake) activity occurs along plate boundaries and subduction zones. The most volatile of these zones is the perimeter of the Pacific Ocean, the infamous Ring of Fire, where the Pacific Plate is disappearing into deep subduction trenches. More than 75 percent of the world's volcanoes are located here, among them such celebrities as Mount Fuji (Japan), Mount Pinatubo (the Philippines), Kilauea (Hawaii), Mount Saint Helens (Washington State), and Popocatépetl (Mexico).

Tectonic plates and the Ring of Fire.

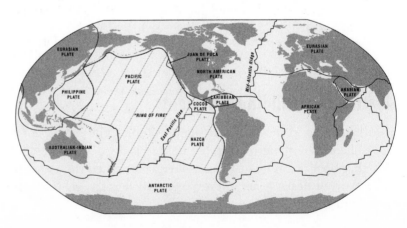

The Great Shake: Earthquakes and Faults

The activity along plates can cause tremendous pressure to build up. In an earthquake, a sudden release of energy causes the rock on one side of the fault to slip rapidly past the other, with either an up-and-down (vertical) or sideways (lateral) motion.

The wave vibrations generated by an earthquake are measured by a device called a seismograph. There are several ways to express the severity of an earthquake. The most common is the Richter scale, a logarithmic scale in which every whole-number increase represents a tenfold increase in earthquake magnitude. A seven on the Richter scale, for example, represents an earthquake ten times as powerful as a six. An earthquake's destructiveness, in terms of loss of life and property damage, depends on its magnitude and the degree to which the area around the earthquake *epicenter* is populated. By the time an earthquake hits seven or eight on the Richter scale, the damage can be catastrophic. A nine on the scale has never been recorded—an earthquake of that magnitude would result in virtually total destruction.

def•i•ni•tion

The spot on the surface right above the focus is known as the earthquake's **epicenter**.

Tsunami or Tidal Wave?

Earthquakes that occur undersea can generate destructive ocean waves called tsunamis. Although tsunamis are sometimes called tidal waves, this is inaccurate because this type of destructive activity has nothing to do with tides. The tsunami is barely noticeable on the ocean's surface; however, once these fast-moving waves reach the shallow water of the shoreline they build up to enormous heights and can cause extensive coastal damage.

On December 26, 2004, a 9.0-magnitude earthquake off the West Coast of the Indonesian island of Sumatra erupted with a force equivalent to 23,000 Hiroshima-type atomic bombs. This violent undersea quake set water racing with the speed of a jet plane across the Indian Ocean. A wall of destructive water hit 11 countries in Southeast Asia, with 150,000 people killed or missing and many millions homeless.

Building Up the Earth's Surface

The movement of tectonic plates is one of the earth's land-building mechanisms. As the continental plates press against one another, the earth's crust folds, warps, and faults. The process of building up the major features of the earth's surface through the movement and deformation of the crustal plates is called diastrophism.

Another primary land-building force is volcanism, which occurs most commonly along the fractured edges of continental plates. Magma, or molten rock, is forced up through cracks in the earth's crust to emerge at the earth's surface as lava. The buildup of lava and ash forms the familiar volcanic cones that soar majestically on the landscape.

Wearing Down the Earth's Surface

Weathering, erosion, and other gradational forces combine to wear away in time even the highest mountains. Prying plant roots and the freeze-thaw cycle eventually reduce huge rocks to small stones. Acids and oxidation (rust) chemically break down the earth's minerals. The wind sculpts sand dunes, while windborne abrasives rasp away at exposed rock surfaces. But of all the agents of erosion at work in the world, the most powerful by far is water.

Follow the Flowing Water

Water erodes, wears down, and carves into the earth's surface in a variety of ways. Rain, tides, flooding, and even ocean waves lapping against rock cause the breaking down and moving of surface features.

Water also moves around great quantities of material. Streams typically originate in steep highlands and mountain areas. As they flow downward, the quickly moving water is capable of moving stones, sand, and silt (the load of the stream). When fast-moving streams and rivers reach the lower, flatter land, the flow begins to slow. The slower-moving water then spreads out to form floodplains, the wide, level valleys that store excess water when the river floods. As these streams and rivers reach the sea, they deposit their loads to form deltas. Deltas are characterized by rich alluvial (water deposited) soils renewed by the continual deposit of silt carried by the river from the land's interior.

The Crushing Effects of Ice

Water can also carve away effectively at the earth's crust in another form: ice. Many landforms in the Northern Hemisphere are the result of glaciation. During the ice ages, the last of which occurred between 8,000 and 15,000 years ago, huge sheets of ice spread southward from the North Pole. These massive 10,000-foot-thick glaciers scoured the land surface and left glacial lakes and deposits in their wake.

The remnants of these continental glaciers are still found today on the island of Greenland and on the continent of Antarctica. Smaller glaciers, called alpine glaciers, are also found in high mountain areas. Alpine glaciers advance and retreat seasonally, grinding the mountain rock into coarse till and a fine mineral dust called glacial flour that causes the spectacular aqua-blue hues of many high mountain lakes.

The Least You Need to Know

- The earth's rotation on its axis once every 24 hours causes day and night.

- The earth's revolution around the sun once a year and the tilt of the earth's axis cause the changing seasons.

- The earth is hottest at the equator because that's where the sun's rays strike the earth most directly. It is coldest at the earth's poles because these areas receive the least direct light from the sun.

- The earth's crust is composed of slowly moving plates that grind and bump against one another.

- The earth's surface is built up primarily by folding, warping (diastrophism), and volcanism; the surface is worn down primarily by running water and ice.

Cultural Geography: The Hand of Humankind

In This Chapter

- ◆ Pick a language—any language!
- ◆ Religious persuasions
- ◆ The distinction between race, ancestry, ethnicity, culture, and nationality
- ◆ The political landscape
- ◆ The business of business

Cultural geography, or human geography, is the study of all that humankind has accomplished and does on this earth.

The human experience incorporates language, belief systems, racial groups, political systems, customs, livelihoods, architecture, economic systems, clothing, and other interactions with the environment. If humankind made it, did it, or believed it, then you have human geography.

Parlez-Vous ... More Than 3,000 Languages?

Thousands of languages are spoken in the world, grouped by linguists into a few language families consisting of languages believed to have had a common historic origin. The largest of these families is Indo-European, which includes languages spoken by about half the world's people.

Within language families are language branches (or subfamilies). The Indo-European family, for example, includes the Germanic (or Teutonic) branch, the Romance branch, and many others. Although languages within each branch might still be mutually unintelligible, they are more closely related than those of a broader language family and more recently diverged from each other.

Branches are further broken down into language groups. Languages within a group share common origins in more recent history as well as a good deal of vocabulary and grammar. English, for example, is a language in the West Germanic group of the Germanic branch of the Indo-European language family.

It's the Same but Different: Dialects, Pidgins, Creoles, and Lingua Francas

A pidgin is a blend of different languages. Not the primary language of any of its speakers, a pidgin evolves so that speakers of different languages can converse. If, over time, the primary languages become lost and the pidgin language becomes the primary language, it's called a creole.

Another means by which people of different tongues converse is in a lingua franca, a language used by peoples whose primary languages are unintelligible to each other. As an example, ambassadors from Iran and Germany might converse in the language of diplomacy—which was traditionally French, but is now English. Catholic priests, by learning Latin, can communicate with other priests across the globe.

Languages, Officially

An official language is one that has been designated by law as the tongue of the country. This type of decree might specify that the language (or languages) be used in schools, courts, businesses, and government. Mastery of the official language might even be a requirement for immigration into the country.

Some countries, like the United States, have not designated an official language, and others, like South Africa with 11, have several official languages. Areas within countries might promote languages different from the official language of the country.

A Look at the World's Religions: The Biggies

Geographers classify religions into two primary types: universalizing and ethnic. A universalizing (or proselytizing) religion is open to all human beings and attempts to spread its faith through missionary activities. (The three largest are Buddhism, Christianity, and Islam.) An ethnic religion generally encompasses specific groups of people in a particular location on Earth. (Hinduism is the largest, and Judaism is another example.) Because they are largely specific to a group or a place, such religions tend not to engage in missionary activity.

Christianity (approximately 2 billion adherents): A type of monotheism, this universalizing religion springs from Judaic roots. Centered on the life and teachings of Jesus, who lived in the first century C.E., Christianity's principal religious text is the Bible, composed of the Old Testament of the Jews and the New Testament. Although this religion originated in the Middle East, today it's widely dispersed throughout much of the world. Major denominations in order of adherents include Roman Catholic (Southern Europe and South America primarily), Protestant (mostly in Northern Europe and North America), Eastern Orthodox (mostly in Eastern Europe), and Anglican (England and North America).

Islam (approximately 1.3 billion adherents): Also a monotheistic, universalizing religion, Islam descends from the Judaic-Christian faiths and is centered on the life and teachings of the prophet Mohammed, who lived in the seventh century C.E. Its principal religious text is the Koran. Islam is the world's second largest religion and the fastest growing. It is practiced worldwide, but is most prevalent in the Middle East, North Africa, and Southeast Asia. Most Muslims (about 85 percent) belong to the Sunni movement, including the stricter form of Wahabi (Saudi Arabia), while about 15 percent belong to the Shiite movement (mostly in Iran, Southern Iraq, and Northern Saudi Arabia). There is also the Sufi mystical tradition.

Hinduism (approximately 900 million adherents): Perhaps the world's oldest religion, the origins of Hinduism date back more than 5,000 years. Linked closely with the people and culture of India, this major religion is the world's largest ethnic religion. Not tied to any one prophet or theological system, it incorporates many doctrines and manifestations of deity. It is somewhat monotheistic yet highly polytheistic in practice. About two thirds of all Hindus belong to the Vaishnavite school and most of the rest are of the Shaivite school, but there are a variety of other schools of Hindu thought, including the 5 million adherents of the related Jain faith.

Buddhism (approximately 375 million adherents): A universalizing religion in the sense that it is open to all, Buddhism is highly tolerant of other beliefs and does not promote preaching or conversion. Buddhism originated as a Hindu reform movement in the sixth century B.C.E., based on the teachings and philosophy of Siddhartha Gautama, known as the Buddha, or "Enlightened One." Although Buddhism arose in India and has adherents around the world, today it is most prevalent in Southeast Asia, East Asia, and Japan. Some forms of Buddhism, such as Tibetan, believe that some of their current teachers are "living buddhas" or gods; others see the Buddha as simply a great teacher. More than half of all Buddhists follow the Mahayana tradition (China, Japan, Korea), almost as many follow the Theravada tradition (Southeast Asia), while about 20 million follow the Lhamsari (Tantric) form (Tibet).

Chinese traditional religion (approximately 394 million adherents): The major Chinese religion is a blending of Mahayana Buddhism, Confucianism, and Taoism, along with local beliefs and practices, and is highly polytheistic. More of an ethical code than a religion, Confucianism was founded in the sixth century B.C.E. by the philosopher Confucius (K'ung Fu-tzu). Taoism also arose in China in the sixth century B.C.E. Its teacher, Lao-tzu, based his philosophy on a simple life in harmony with the way of nature. Chinese traditional religion is practiced primarily in China, Korea, and Southeast Asia.

Judaism (approximately 14 million adherents): The first monotheistic faith is rooted in the teachings of Abraham and the Torah of Moses. Judaism is thought to be between 3,000 and 4,000 years old. Most of the world's Jews today live in the United States, Israel, and Europe, but there are also large Jewish communities in Canada, Latin America, and South Africa.

In total numbers, about 33 percent of the world's people follow the Christian faith, while 21 percent are Muslims, 14 percent are Hindu, 16 percent are nonreligious, 6 percent follow Chinese traditional folk religions, and 6 percent are Buddhist. The balance of humankind includes .4 percent Sikh, .2 percent Jewish, and the remainder, including indigenous and other religions.

Race, Ancestry, Ethnicity, Culture, or Nationality— What Does It All Mean?

Although the terms culture, ethnicity, and race are often used somewhat interchangeably, they have specific meanings to geographers.

The Diversity of Race

A race is an identifiably distinct group of people, whose distinctiveness is based on inherited biological characteristics. These distinguishing characteristics include, among others, hair texture, height, facial characteristics, skin pigmentation, and even variations in blood type, each of which typically represents a physical response to the environment evolving over long periods of time.

The distinctions between racial groups can be extremely subtle, however, and variations within a racial group can be extremely wide. Although other geographers' lists might vary somewhat, these are the nine generally recognized racial groups:

- African

- Asian

- Australian (aboriginal, not "Crocodile" Dundee)

- Caucasian

- Indian (from South Asia, not American Indian)

- Indigenous American

- Melanesian

- Polynesian

- Micronesian (not always included in this list)

Ancestry (The Family Tree) and Nationality

The term ancestry simply refers to one's line of descent, or lineage.

Whereas nationality is often confused with ancestry or even race, it is actually more of a legal concept, referring to a person's membership in a particular nation, either by birth or acquired citizenship. A nation can be thought of as a unified group of people sharing customs and ethnicity and usually occupying a territory, but not necessarily a state, which is a political entity. Unlike race or ancestry, it's possible to change nationality by becoming a citizen of another country (or state). The process of acquiring citizenship in a country other than the one in which you were born is called naturalization.

Coping with Changing Culture

Culture is a broad term that can be used to describe a society's entire way of life—its pursuits, behaviors, ideals, values, pleasures, and dreams. Technology, art, food, fashion, symbols, music, belief systems, and laws are all aspects of culture. As these elements are passed down from generation to generation and absorb outside influences, a culture becomes richer and more complex.

def•i•ni•tion

Diffusion is the spread of ideas, cultural items, or knowledge from their origins to areas where they are adopted.

Culture is *diffused* from place to place in a number of ways. In expansion diffusion, a culture trait (item or idea) stays strong where it originated (its culture hearth) but moves out from that area into neighboring areas. Hierarchical diffusion has people copying the ideas or physical objects of the rich and famous, or the leaders. Relocation diffusion sees the idea/item die out in its place of origin (culture hearth) but relocate or become viable in some distant place. Stimulus diffusion sees an item/idea moving from its culture hearth and becoming something different in a new place, but modeled after the original.

When two cultures meet, the weaker of the two often takes on aspects of the stronger, which is called acculturation. When a culture totally submits and becomes just like the dominant culture, it is called assimilation. In rare cases, like Mexico, two cultures have come together and blended into a new culture; this is called transculturation.

An Eye on Ethnicity

A person's ethnicity can be based on many different criteria, such as racial characteristics, language (or dialects), religion, nationality, and culture. Any of these factors can identify an ethnic group as distinct from a larger population.

An example of the difficulties that can arise in defining ethnicity is the frequent misuse of the term Hispanic in the United States. The term is often listed in population tables as a race, although people of Hispanic origin can be of any race. Nor does it refer to any one nationality. Hispanics might be from Mexico, Puerto Rico, Cuba, the Dominican Republic, Central America, or South America, or a number of other countries. The common characteristic shared by people who describe themselves as Hispanic is either the use of Spanish as a primary or secondary language or family origins in a Spanish-speaking country.

Terms of Settlement: The Political Landscape

Geographers often use the term political to describe maps that feature human-made boundary lines and divisions, as opposed to physical maps that show mountains, rivers, and other features. Countries, states, territories, counties, and cities have defined (although sometimes contested) outlines that appear on a map but not on the actual physical landscape.

Here are some of the ways geographers define boundaries:

♦ **Natural versus artificial boundaries:** Boundaries based on landscape features (such as mountains, lakes, and rivers) are called natural (or physical) boundaries. The boundary between Texas and Mexico, for example, follows the Rio Grande and is therefore a physical boundary with lots of curves and wandering. If you see straight lines dividing places on a map, you're probably looking at geometric (or artificial) boundaries, which are usually surveyed lines often based on a parallel or meridian (a line of latitude or longitude). The straight boundary between Arizona and New Mexico is an example of an artificial boundary.

 Terra-Trivia

Not all boundaries are exact. The Karakoram and Himalaya mountains are a boundary nightmare. Jagged peaks, violent storms, shifting glaciers, and avalanches all cause regular border disputes among Pakistan, India, and China as they lay claim to this region.

♦ **Antecedent versus subsequent boundaries:** Boundary lines drawn before an area has been well populated are antecedent boundaries. The western border between the United States and Canada is an antecedent border, drawn before "the West was won." (Native Americans were not factored into the governments' decision.) Borders drawn after an area has developed cultural characteristics are called subsequent borders, such as new boundaries drawn after the breakup of the former Yugoslavia.

♦ **Consequent versus superimposed subsequent boundaries:** Boundaries drawn after an area has well-established cultural characteristics can either respond to those characteristics or ignore them. Consequent (or ethnographic) borders are drawn in an attempt to respond to cultural dictates, such as what occurred when the Czechs and the Slovaks decided to go their own ways as separate nations in 1993. Superimposed boundaries are borders drawn rather arrogantly that "do their own thing" regardless of any preexisting cultural patterns. Superimposed boundaries placed by colonizing European powers, for example, typically divided indigenous peoples. Africa is laced with such boundaries. Tribal areas were simply cut in half as European countries divvied up the land of Africa by imposing boundary lines on a map.

Countries, Nations, and States

The distinction between countries, nations, and states can be somewhat confusing. First, forget about the use of the word "state" to define the political units that form a federal government (such as New York State or the state of California). What concerns you here is the use of the word "state" to describe a country or nation:

◆ **State:** An independent political unit with an established territory, boundaries, population, and government controlling its own internal and foreign affairs.

◆ **Country:** A state, or the territory of a state. (The word "state" has more of a constitutional and political connotation; "country" has more of a territorial connotation.)

◆ **Nation:** A closely linked group of people occupying a territorial area.

Stateless nations do exist. The most visible examples are the Palestinians of the Middle East and the Kurds, who live on lands straddling Turkey, Syria, and Iraq. In a nation-state, a distinct nation of people can be the primary occupants of a state (Italy, France, and Japan, for example). Many examples exist of multinational states in which two or more nations of people occupy one state (the Flemish and Walloons in Belgium, for example).

Sorting Through Settlements

The people of the earth are becoming more urbanized. As global economies shift away from an agricultural base, more and more people are flocking to urban areas. Many of these areas are growing at an alarming rate, especially in the developing world. About 40 percent of the population in developing countries is now urban; that figure is expected to increase to 56 percent by 2030. Urban refers to a centralized area regardless of size.

The following terms describe urban areas, in order from smallest to largest:

- **Hamlet:** Tiny; maybe just one or two dozen buildings, usually with no stores or businesses.

- **Village:** Larger than a hamlet but not as large as a town, usually with a "general store" or small-scale business.

- **Town:** Bigger than a village and generally centered on a nucleated business area.

- **City:** Same concept as a town, only larger and more complex; usually containing many nucleated business areas.

- **Conurbation:** A large, built-up area in which two or more cities have coalesced into one huge unit (think the Tokyo-Yokohama-Kawasaki conurbation).

- **Megalopolis:** A huge, built-up area in which several conurbations have coalesced; the Boston-to-Washington corridor on the East Coast of the United States is a prime example.

Getting Down to Business

As you surely already know, vast economic differences mark the countries of the world. One gauge of economic status is per capita gross domestic product (or GDP), which is how much money a country produces per person living in the country. Countries such as Somalia (in Africa) have a per capita GDP of $600. At the other end of the scale are countries such as Bermuda at $69,900, while the United States comes in at $41,600.

Although the economies of most countries are blends of two or more types, there are still only three basic types of economic systems: subsistence, commercial, and planned.

A subsistence economy is based in rural areas and around family units. In this survival economy, people produce goods and services for their own consumption and that of their immediate family. The exchange of currency and goods is minimal, and people are often just one natural disaster away from famine. Typical of this economic type is slash-and-burn agriculture nomadic herding and intensive small-plot agriculture. Subsistence economies are prevalent in the world's least developed countries.

In a commercial economy, the most widespread in the world, people freely produce goods and services for sale in competitive markets, where prices are determined by the law of supply and demand. This is the dominant type of economy in developed countries and has been making rapid inroads in developing countries as well.

In a planned economy, the central government controls the production and distribution of goods and services and sets prices. Mainly associated with communist governments, planned economies are notoriously inefficient. With the dissolution of the Soviet Union and the collapse of Eastern European communist regimes in the early 1990s, the number of planned economies around the world has declined dramatically.

The Global Economy

Globalization has to do with the freer, more rapid movement of goods, services, labor, technology, and capital around the world. This phenomenon is not new, but advances in technology and telecommunications have greatly accelerated it.

Globalization and the lowering of trade barriers have helped raise growth rates and living standards in many developing countries. China, still a communist country politically, has embraced market capitalism with gusto and is now the world's second largest economy. Economic liberalization in India continues to fuel robust economic growth. In Latin America, on the other hand, globalization is being blamed at least in part for the renewed political instability and financial woes plaguing much of the region.

There's more to globalization, however, than its purely economic aspect. The globalized world is a much smaller and more interconnected place, not only economically, but also culturally, socially, politically, and environmentally.

The Least You Need to Know

- Although thousands of languages are spoken on the earth, they are classified into just a few language families.

- Christianity, Islam, Hinduism, Buddhism, and Chinese traditional religion are practiced by more than 75 percent of the world's population.

- The three major types of economies are subsistence, commercial, and planned.

- Globalization is bringing down barriers between nations and making the world a much smaller and more interconnected place.

Chapter 4

Factors in Human Geography

In This Chapter

- ◆ Nomads, pastoralists, and agriculture
- ◆ Cultural diffusion
- ◆ Getting along with others …
- ◆ Cavemen to computers—a changing world

The human species has come a long way from the primitive hunter-gatherer. Yet, with all that people have accomplished, are we that different from the primitive hominid that first walked upright?

Migration—It's Not Just for Birds

Thousands of years ago, before mankind developed civilizations and empires, people were nomads, seeking food and shelter wherever they could be found. The basic hunter-gatherer lifestyle changed when humans began to make tools and hunting began to provide part of the diet, and changed even more once animals were domesticated, allowing people to produce food from their flocks. Through all those ages, however, humans were nomadic and their activity space (the amount of territory used by a specific group of people during their regular activities) was limited.

Although these primitive gatherers, hunters, and shepherds traveled far and wide to find food and safety, they used very little of the earth's surface, and man's imprint on the planet was minimal.

Agriculture, the Beginnings of City Life …

Once people learned to grow food crops (the first was probably wheat or barley), things began to change. The world's first permanent settlement was at Jericho, in the Middle East, and this began a process which grew into the four great earliest civilizations—Egypt, on the Nile River; the Fertile Crescent, along the Tigris and Euphrates Rivers; India, on the Indus River; and China, which began on the Yellow River. This was the beginning of the ecumene (parts of Earth's surface permanently settled by people), and although most of the planet is in the nonecumene, or parts not permanently settled by man, including deserts, mountains, oceans, frozen tundra, and other areas humanity has not yet settled, the ecumene has grown by leaps and bounds throughout human history.

As people settled into the more desirable places, humankind eventually exceeded the carrying capacity, or maximum population an area can support without experiencing unacceptable deterioration in the way of food, water, shelter, or other quality of life issues. This caused internal conflict, the need for some people to move away, or, too frequently, wars with neighboring peoples for resources.

Culture and Ethnicity

Because each group of mankind developed unique cultural identities (language, religion, clothing, etc.), genetic predispositions (race, height, features), or some other discernable feature, mankind divided into varied ethnic groups. (Often, these groups were ethnocentric, each believing that its group was superior to all others.)

Each area of the world, and each group of people, formed distinct cultures in keeping with their peculiar environmental restraints or advantages. The culture-environment tradition refers to this relationship between human societies and their practices as they relate to their natural environment. In every part of the world, mankind works to tame the environment and has come up with ingenious methods of surviving and prospering.

Some scientists ascribe to the theory of environmental determinism, which states that the environment determines the culture and that people can only accomplish what the particular environment allows. Others ascribe to the concept of possibilism, which maintains that humans decide their culture, not the environment; people can overcome any challenge set by the environment of a place, like today's water parks in the desert of Las Vegas, Nevada.

Hearth and Home

We call the place of origin of an ethnic group (or an idea, language, religion, etc.) a culture hearth. Each culture hearth has distinct culture traits (a single element or practice in a culture). Sometimes, a group of culture hearths are interrelated, sharing similar culture systems. For example, the United States and Canada share a common culture, called a culture realm. An area with vast differences, but sharing some common traits, is called a culture region. You can find chopsticks throughout East Asia, for example.

Not all ideas are equal, and various culture groups zealously guard their treasured ideals against outside intervention. The time-distance decay model states that there is a lessening of acceptance of an idea or invention with the increase of time or distance from its origin or hearth. One's own culture is so strongly valued that frequently we find examples of cultural lag, or the retention of culture traits or traditions that are out of keeping with the times or changed circumstances of a society (for example, keeping traditional agricultural practices despite the fact that new technology makes them obsolete).

Cultures have developed all over the world and, sometimes, they have come up with similar solutions or traditions. Multilinear evolution states that parallel but independent cultural development explains why widely separated peoples existing in similar environments, but who could not have interacted with each other, came up with similar cultural or technological solutions.

This ethnic reality played an important part in the development of humankind over the past 6,000 years. Because human populations continued to grow, often exponentially, the communities could not keep abreast with feeding and giving shelter to everyone, so modern man's history is full of constant migration (movement from one place to another). This also entailed adapting, changing traditions, and cultural interactions, both positive and negative. Although primitive man was a nomad, moving from place to place almost daily, modern mankind's story is one of moving from settled, overpopulated areas to new or better locales.

The pressures exerted by population in relation to land are calculated in a number of ways:

- Agricultural density is the number of rural people in an area divided by the amount of arable land (the number of "farmers" compared to farmland), as in 10 per acre or hectare.

- Physiological density refers to the total number of people (city dwellers and farmers) compared to the amount of arable land (usually a much higher number than the agricultural density for a country).

- Arithmetic population density is the total population of a country divided by the total land (including desert, mountain, and lakebeds) of the country. The number of people per unit of land using this calculation is always smaller than the number you get using agricultural or physiological densities.

There may come a point when the planet cannot support all the vast numbers of humankind, because food supplies may run out as a result of our population explosion, or exponential growth, in the past 300 years.

There is also pressure on a population to support all its members. Ancient civilizations could not afford a large "leisure" class. Most people had to work and contribute to the food production, defense, or economic development. The dependency ratio refers to this problem in that it estimates the number of dependents (children, disabled, and seniors) that 100 people in their economically productive years must support. Or, more precisely, the number of dependents a group of 100 workers can support without degrading the social well-being.

Diffusion and Culture

This movement of people from place to place brought with it cultural diffusion, or the spread of cultures to new places. It may take many different forms:

- Contagious diffusion is when an idea or item is passed from person to person or place to place through direct contact, but does not spread uniformly to all people in the area.

- Expansion diffusion is when the idea or item originates in a hearth and spreads outward in all directions from that hub while remaining strong in the hearth area.

◆ Relocation diffusion occurs when an idea or item is completely relocated from its hearth to some new place, like the Mormon Church migration from New York, through Illinois, and finally relocating to Utah.

◆ Hierarchical diffusion involves "followers" or fans adopting the ideas, manners, or cultural items of an upper class, as when people adopt the style of their favorite movie star.

Diffusion and culture are highly involved in religious belief systems. Although we often like to think that our beliefs are unique, or have been given to us directly from "the highest being," we find that the pattern of human belief is fairly consistent and follows some basic formats.

Many ancient religions, in all parts of the globe, seem to have started out as animism. Animism is the belief that objects in nature, such as the wind, rivers, rocks, trees, or animals, are divine or have souls and can help or hinder human activity. Modern examples include the Shinto religion of Japan.

Animism leads to a form of polytheism, the belief system that includes many gods. The ancient Greeks are one example, yet as time went on the varied Greek deities began to merge into the idea of some sort of "universal power(s)." Perhaps this is where the concept of monotheism (the belief in one all-powerful God) began to develop.

Can't I Just Buy an Airplane Ticket?

Migration involves a migrant, or person who moves from one location to another. Those leaving their homeland, or cultural hearth, are called emigrants, while those coming into a new place are called immigrants. They may move in channelized migration, which involves the flow of people between areas that are socially or economically allied in some way (like the early colonists coming to the United States from Britain); or in cluster (chain) migration, in which they may be drawn to an area because their relatives or countrymen have already migrated there, like the Swedes to Wisconsin and Jewish people to New York City. Sometimes, a group of people immigrate, only to return to their homeland; this is called counter migration.

All migrants face hard choices. Sometimes intervening opportunity determines their final destination. For instance, a woman leaves her village in China hoping to migrate to the United States, but when she makes it as far as Hong Kong, she finds a good job and ends up settling there. This is especially a consideration in step migration, where a person cannot go directly to a desired destination, but must make a number of stops, often taking years at each stop, before making it to the goal.

An important factor in migration is critical distance. This is the distance beyond which cost or effort determines people's willingness to travel to a location. Distance bias states that short journeys are favored over more distant ones. (It also means that people are more likely to shop near home.) Friction of distance is the measure of the retarding effect of distance on spatial interaction.

Another complicating factor for immigrants is the connectivity of two locations. This is the amount of direct linkage between one location and other locations in a transportation network. (Orlando has good air connectivity with New York City due to 45 direct flights daily, but poor connectivity with St. Louis, because one has to change planes through Atlanta.) The ability of migrants to go from one place to another is often calculated on a gravity model, a mathematical prediction tool for understanding the interactions of places in terms of population size and distance between them (how a housing slump in Florida affects migration from other parts of the country, for example).

The gravity model is one facet of the global-local continuum that explains that what happens at the global level has a direct effect on what happens at the local level and vice versa. In other words, the world is a web of interconnected relationships.

Can We Get Along?

When people migrate from one place to another there are always difficulties, especially in today's crowded world, because they come into frequent contact with other ethnic groups, ideas, and cultures. All ethnicities want to maintain their uniqueness, but they also have to fit into a new cultural reality. Will they assimilate? Can people give up their former culture and become a full part of their new homeland's culture? This is the goal of the melting pot model in the United States; from many peoples and cultures we have one "American" culture. Also called amalgamation, the melting pot is a merging of the culture traits, ideas, and lives of all their member cultures and peoples. Although the United States started as a "British" colony, it has adapted and accepted cultures and peoples from around the world.

Yet other societies have attempted to live in a state of cultural pluralism where two or more population groups (religious, ethnic, linguistic, etc.), each practicing in their own unique manner, live adjacent to each other without mixing. An example of cultural pluralism is the country of Belgium, where Flemish and Walloon live in uneasy coexistence. Belgium is also a multicore state made up of more than one core or dominant religious, linguistic, economic, or cultural group working in harmony within one governmental framework.

Some groups have gone through a process of acculturation, where they adopt outside technology and ideas after coming into contact with a more technologically advanced culture, such as Central Africa's borrowing from Europeans in the 1700s. Even rarer is the concept of transculturation, where two equally advanced cultures come into contact and mutually share to form a brand new culture merging the two. The Mexican culture was formed when the Spanish interacted with the Amerindians in this transculturational process.

No matter what form the transition takes, it is always difficult for an immigrant to fit into the new place. Even in a melting-pot society like the United States, immigrants face a daunting number of difficulties and, often, discrimination. Humankind seems to have a built-in "we" versus "they" mentality. It's not just ethnic groups; just try to get a University of Florida "Gator" to coexist with a Florida State "Seminole"!

Forces that tend to divide a country—often religion, economics, language, or ethnicity—are called centrifugal forces; those issues that unite and bind a country's people together, often the very same issues, are called centripetal forces. A rare example of nations working together for the common good is the European Union. This recent addition to the world scene has the nations of Europe bonding together in a loose federation for economic benefit. This is a good example of supranationalism, or a group of national states working together as a team, such as NATO, the EU, and NAFTA. This is the very opposite of supernationalism, or the goal of a nation to dominate others because they perceive their state to be "superior" in some way, such as the Nazis, or even British imperialism of the previous era.

Of course, migration also depends upon opportunities—the lack thereof in the homeland, and the possibility of opportunity in the hoped-for new home. People in poor, underdeveloped countries often rely on slash-and-burn, subsistence agriculture (also called milpa agriculture, swidden agriculture, or shifting agriculture) to barely survive. For such people, the possibility of a better life is a big inducement for migration, whether to another location in their homeland or to a new homeland. The problem is that every area has a maximum sustainable yield, the maximum rate at which a renewable resource can be exploited without impairing its ability to renew or replenish itself. Milpa agriculture perpetuates itself, and keeps people poor, because the land does not improve and they are forced to constantly find new croplands or other means of sustenance.

The tragedy of the commons is an interesting concept that states that in the absence of government control over the use of a resource, the individual will maximize his or her own individual share, even though this collective plundering may diminish the yield or destroy the resource altogether. Perhaps this explains why immigrants are often mistreated and exploited, why even the United States has virtually ignored the countless illegal immigrants in the country: because the immigrants will work "cheap."

Borders and Other Barriers to Migration

For some reason, humans have a nasty habit of claiming ownership of things. Individuals, ethnic groups, and countries all seem to want to own things. Of course, this usually involves the notion that "others" may not use these things, and it frequently means that one group controls another.

Boundary is a construct of man. It demarcates between one entity and another, and comes in many forms:

- An antecedent boundary is one that was established before an area became populated or belonged to the current cultural group.

- A subsequent boundary is established after the area in question has been settled, such as the border between Viet Nam and China.

- A consequent boundary (a type of subsequent boundary) refers to a boundary line that coincides with some cultural divide, such as religion, language, or ethnicity.

- A superimposed boundary is one placed over and ignoring an existing cultural pattern, such as the line separating the two halves of New Guinea.

- A geometric boundary, sometimes called an artificial boundary, is without obvious physical basis in the natural landscape; rather, it is an artificial straight line dictated by political necessity.

- A natural boundary is based on a recognizable geographic feature, such as a river or mountain range.

- A relic boundary is still discernible and marked by some cultural landscape feature, even if it has no current political or social meaning, like the Berlin Wall.

- An isogloss is the boundary line marking the limits of a particular linguistic feature, language, or dialect.

Governments

In ancient times migration was most hindered by geographic barriers. Today the greatest challenges for migration are political. Since the beginning of time, people have attempted to control others. In ancient times, and even today, a common form of government was an autocracy, or one-man rule. An autocrat might be a good ruler or a bad one. The ancient Greeks had tyrants, who ruled with the consent of the people. The Romans would appoint a dictator to rule for a period of time during a national emergency, then put the dictator on trial after his term of office if he misbehaved. Today we have no such restraints; the tyrants and dictators of today are harsh, self-serving, one-man rulers.

Of course, the ancient Greeks also gave us the concept of democracy, or rule by the people. Every citizen got together with all other citizens and voted on every issue. This worked well in ancient Athens, where the population numbered only a few thousand citizens, but becomes problematic in the modern world. As a result, most "democracies" of today are actually representative democracies, where all the citizens vote for representatives who actually run the government and make the laws.

In an oligarchy a group of people rule as a committee. Oligarchs are usually important men of wealth or education, or they're outstanding in some other way. Ancient Sparta had an oligarchy of the ephors; these five men were the oldest, most distinguished military men in their city-state. A junta is a military oligarchy; usually such a ruler gets his job by overthrowing the previous government with military might. An aristocracy is an oligarchy of birth. Each noble, or aristocrat, came to power because his or her parent was an aristocrat.

A monarchy is similar to an aristocracy, and they often work together, but in a monarchy all power is in the hands of the hereditary ruler—the king, queen, shah, sultan, emperor, maharajah, prince, etc. In a theocracy the government is ruled by religious leaders who claim authority as their god-given right.

In addition, a government may be a federal state in which the central government of the state has specific powers, but the individual regions (states in the U.S. federal system) retain certain powers to make and enforce their own laws, policies, and customs. In a unitary state the central government retains all power over all parts of the state.

No matter what kind of state one lives in, nor what form of government rules, there is a factor in law known as distance decay. This is the degenerative effect of distance on human spatial structures and interactions, as when people were less law abiding when far from government agencies in the 1800s United States "Wild West." This also relates to another concept, the law of peripheral neglect, which states that the further an area or issue is from the central government the less likely the central government will be concerned or deal with it in a timely fashion.

States

To be a sovereign nation (country), every state must control some land. Therein lies the problem, because no state ever seems to have enough land, resources, or people to control. This has often led to wars, untold misery, and lots of legislation.

States usually make cadastral maps, which show the value, extent, and ownership of land for the purpose of taxation. They also try to control and tax basic activities (economic activities involved in production and exportation from a country) and nonbasic activities, or service industry, which involves the production and consumption of products within the state.

States come in a variety of shapes and sizes:

◆ A compact state has a roughly circular or rectangular shape in which all points on the boundary are roughly the same distance from the center; examples include Poland, Uruguay, and Cambodia.

◆ A prorupted state has a basic compact shape with an additional "arm" or other projection that juts out from the norm, like Thailand. Chile is an example of a long and narrow, or elongated state.

◆ A perforated state is one whose territory completely surrounds that of another, such as Italy in regard to San Marino.

◆ An enclave is a piece of territory that is surrounded by the territory of another state. The various lands of the Vatican are enclaves in Italy.

Terra-Trivia

Mongolia is the least crowded country in the world with only 4.3 people per square mile. Namibia follows with only 5.7 and Australia with 6.6 people per square mile. Of the larger, developed nations, Canada has only 9.5 people per square mile.

◆ An exclave is a non-island piece of territory that is part of one state but is separated from the rest of the state by the territory of another state. Alaska is an exclave of the United States, because it is separated from the "lower 48" by Canada.

◆ The United States is also a fragmented state, one with parts separated from the main part, because of the exclave of Alaska and the Hawaiian Islands.

For various reasons states frequently have regions or groups that demand and gain some degree of autonomy or even independence from the central government. This process is called devolution. Balkanization is an extreme form of devolution where a country fragments in smaller, often hostile, areas, as in the breakup of the former Yugoslavia.

Cavemen to Computers

Since humans first began to settle and build civilizations, they have been divided by the ways in which they earned their living. Early on there were the gatherers, then the more elite hunters, followed by pastoralists, and then farmers. Each was a distinct advancement of primitive man to greater "civilization," and new possibilities came about with each innovation and discovery.

Basically, we can today divide human economic activity into five areas:

◆ Primary economic activity is concerned with direct extraction of natural resources, including fishing, farming, herding, mining, lumbering, and hunting.

◆ Secondary economic activity involves the processing of raw materials, or manufacturing.

◆ Tertiary economic activity refers to the service industry, such as office jobs, retailing, education, banking, and service jobs such as waiting tables and caring for lawns.

◆ Quaternary economic activity refers to the collection, processing, and manipulating of information and capital, as in the insurance, legal, and finance realms.

◆ Quinary economic activity refers to any highly technological industry such as scientific research or high-level management.

Today most people in developed ("ped") countries, such as the United States and Canada, most of Western Europe, and Japan, work in the tertiary, quaternary, and quinary sectors. Very few people in those developed lands work at producing food or securing raw materials. The opposite is true in the developing ("ping") countries, where the bulk of the population is involved in producing food, often at the subsistence level.

This all relates to the continuing growth of the world's population and the shift from agricultural to manufacturing, and now service industry in the developed world. This is part of the demographic transition model, based on Western Europe's changes in population growth during industrialization. High birth rates and death rates are followed by plunging death rates, which produce a huge net population gain. This is followed by the convergence of birth rates and death rates at a low overall level. Demography is the study of rates of population change, including birth and death rates, migration, and population distribution. Demographic variables include fertility rates, mortality rates, and migration patterns.

Thomas Malthus, a nineteenth-century thinker, proposed that the population growth was outrunning the earth's capacity to produce sufficient food. He predicted dire consequences, including massive wars over food. The anti-Malthusian school counters with the theory that science will prevail and humankind will, through the technology of better foods and fertilizers, keep food production in sync with the population. Neo-Malthusians speculate that population-control programs to preserve and improve the quality of life would save humankind and keep us in sync with our food limitations.

Commodification: Everything Has Value?

Commodification is the process through which something is given monetary value. This is when an item or idea that was previously deemed worthless becomes desirable. (Pet rocks were just rocks until somebody marketed them and people bought them.)

Commodification is the basis of our modern economy. In other words, diamonds and gold would be valueless if people did not consider them valuable. Swamp land in Florida becomes valuable real estate when somebody wants to build houses on it.

This basic idea has brought about our modern world's economy. Things have value!

When something has value a sort of chain reaction begins. This is called the multiplier effect, in which the expansion of one economic activity produces other economic activity. Opening a new shoe factory not only opens up jobs in the factory, but also allows other businesses, perhaps a restaurant or gasoline station catering to the new factory workers, to enjoy economic activity and profit, and create other jobs in the area.

Once a business begins, once economic activity is initiated, man has always sought better and more efficient ways to gain more economic success—to become richer, in whatever is deemed valuable in that society or time. The least-cost theory states that the location of a manufacturing center is determined by the costs and advantages of labor, transportation, and agglomeration or deglomeration. In other words, a successful business depends on the three factors of labor: cost and availability, transportation (to bring in raw materials and export finished products), and proximity of other businesses or support industries.

This is further delineated in the idea of economies of scale. This is the savings that accrue from large-scale production of goods because the unit cost of manufacturing decreases. In other words, big businesses can buy raw materials cheaper, in bulk, and save money on production costs, thereby getting a competitive edge on smaller businesses. (The old corner grocery is giving way to super supermarkets, because customers will give up the "comfort" of their little grocer for the cheaper prices at the huge "warehouse" store.)

Of great importance to this economic issue is the concept of agglomeration. There are benefits for businesses of a specific type to put their factories in close proximity to each other in order to share labor, technology, supplies, etc. The agglomeration of car factories in Detroit also brought clusters of subsidiary manufacturers of headlights, tires, upholstery, etc. Deglomeration is the opposite. It is the process of deconcentration of an industry in response to advances in technology or rising costs due to congestion and competition.

In general, things in the economic world do not just happen. Things are planned so that the greatest possible benefits and profits may be obtained. Even the simple shipping of goods from one place to another has become something of a science.

Singapore has grown up as an economic power simply because it is fortunate enough to sit in the center of major ocean shipping routes between major producers and markets. It serves as a world-leading entrepôt, also known as a port city or major shipping hub where goods are stored, imported, or transshipped (it's also called a break-of-bulk point). A huge ship comes to Singapore from mainland China filled with radios. The radios are unloaded, and then reloaded into other ships, with varied cargo, going to other ports around the world. It saves a ship having to go to dozens of ports around region to get a load of varied goods.

This also explains why, all over the world, a basic pattern of economic growth prevails. Location theory explains the interrelatedness of economic activity. We put factories near power or energy sources and raw materials; we put malls near major population centers; and we see major suburbs grow up along major access routes between centers of commerce and industry.

Although many older cities grew outward from a central business district, there is a more modern approach called the multiple nuclei model, in which the modern American city does not necessarily have a distinct "downtown" area; rather, today's U.S. cities have multiple "centers." The urban realms model reflects the frequent reality of the late-twentieth-century U.S. city as a widely dispersed metropolis with many centers. Each center becomes a basic "downtown" for that area of the city. This is the opposite of the traditional U.S. city, which is totally focused on a Central Business District to which all suburban and urban activity is directed.

Many regions of the world, or a country, are known for what they produce. This is called a functional region, or an area defined by a specific function, product, or activity. (Think the "steel belt" around Pittsburgh, or the auto industry in Detroit.) Other regions might be called formal regions, or uniform regions, because they have a high degree of homogeneity in one or more phenomena. (The "old" South fits this description, with its fried chicken, cotton crops, pecan pie, and iced tea.)

Today, if we are to believe world-systems analyst Immanuel Wallerstein, the world is one giant interconnected whole. Social change in the developing world is linked to the economic activities of the developed world. In this theory the world functions as one organic entity organized around a new international division of labor in which workers in the less developed countries have no recourse but to serve their developed "masters." This accounts as a new "colonialism."

The Least You Need to Know

- Mankind has always moved around.
- Mankind has invented many ways to try getting along with others.
- Everything has value, when people "own" things.
- Human geography uses many "models" to explain human economies.

Chapter 5

Maps: A Geographic Love Affair

In This Chapter

- ◆ Identify different types of maps
- ◆ Know why a map can never be perfect
- ◆ Let symbols on the legend help you read a map
- ◆ Use a map to locate positions and measure distances
- ◆ Read the ups and downs of the earth's surface
- ◆ Develop a sense of direction

Maps are the basic tools of geography. Maps provide an enormous amount and variety of information, including relative size of a place, distances between places, directions, and statistics about a place.

Millions of Maps

Maps exist for virtually every type of information under (and including) the sun. One way to categorize maps is by the type of information they convey. Here are a few common map types:

◆ **Weather:** Often on TV news and in newspapers, these maps help people plan their activities.

◆ **Road:** Indicating types of roads, relative distance, and other vital information for drivers.

◆ **Political:** Shows the boundaries of nations and their political subdivisions (states or provinces, counties, cites, townships, and so on).

◆ **Physical:** Presents the earth's physical features: mountains, valleys, deserts, rivers, lakes, and so on.

◆ **Coastal chart:** Extremely useful for boaters and fishermen.

◆ **Star chart:** A map of the heavens.

Pondering the Projection Dilemma

A map represents the ins and outs of the earth's surface on a flat piece of paper. If you've ever given a friend directions by scrawling a map on the back of an envelope, you know the basic process. Mapmakers (cartographers) attempt to do the same thing for the entire earth. Although this process might not seem difficult, there's a fly in the proverbial cartographic ointment.

When you peel an orange and attempt to lay the peel flat on a table, you begin to see the problem cartographers face: the earth is not flat. The spherical earth—an oblate spheroid, to be exact—poses a fundamental dilemma to cartographers: no matter how hard a mapmaker tries, the rounded surface can't be represented with 100 percent accuracy on a flat map.

Creating Projections

Cartographers call their attempts to create flat representations of the spherical earth map projections. Creating a perfect projection is impossible.

The only truly accurate representation of the earth is a globe because it preserves all the map characteristics of interest to a cartographer: area, shape, direction, and distance. A map projection can preserve only one or two of these characteristics. If you focus on getting an area's size exact, for example, its shape might be distorted, and vice versa. Cartographers have names for projections that preserve each of these characteristics:

- ◆ **Equal area:** Area
- ◆ **Conformal:** Shape
- ◆ **Azimuthal:** Direction
- ◆ **Equidistant:** Distance

Mercator's Famous Projection

One of the most important early map makers was Gerardus Mercator, a sixteenth-century Flemish (Belgian) geographer, mathematician, and cartographer. Mercator's projection was influential because it preserved shape and direction, allowing sailors for the first time to use a map to plot "true" direction. The problem with the Mercator projection is that it badly distorts area, especially toward the poles. Eurasia and North America loom very large in contrast to South America and Africa. Despite this basic flaw, the Mercator projection stuck for centuries, although it has been largely superseded today.

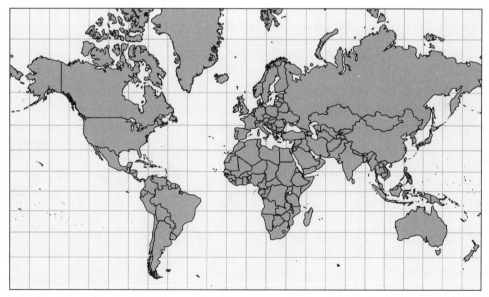

Mercator's projection.

Terra-Trivia _____

One of Mercator's other claims to fame was that he published the first bound collection of maps to be called an atlas.

Sorting Through the Symbols

Because maps represent a reduced part of the earth's surface on a small piece of paper, certain symbols (points, lines, and patterns) are often used to represent landscape features. Cartographers also use color to help map users "read" a map:

◆ **Blue:** Water features (rivers, lakes, and ponds, for example)

◆ **Brown:** Height and depth (mountains and valleys, for example)

◆ **Green:** Vegetation features (woods, orchards, and scrub brush, for example)

◆ **Red:** Roads and urban development (important roads and built-up urban areas, for example)

When a map has many symbols or its symbols are tricky, it usually includes a legend (sometimes referred to as a map key) for explanation. See the following example of a legend.

A map legend.

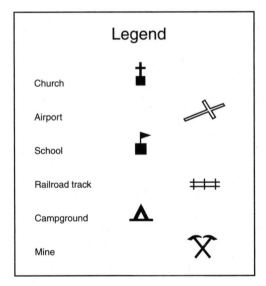

Now, Where Was I?

You can locate a position on a map in two basic ways: *relative location* and *absolute location*.

Relative Location

You generally use relative location when you're giving someone directions. To direct a person to the local convenience store, you might say something like this: "Go to the airport, and continue through three lights; the Trendy Mart is on your right." These directions give the position of the Trendy Mart relative to the airport.

def•i•ni•tion

Relative location specifies a particular position or place by relating it to another known position or place. **Absolute location** indicates position based on the coordinates of a grid system (such as latitude and longitude).

Absolute Location

Absolute location identifies the position of a place by means of a grid system. Although absolute location is a useful way to locate something in an atlas, it's not often used in everyday speech. The person who wants to find Trendy Mart will have a hard time finding it if you say, "Proceed on a 232-degree azimuth to coordinates 41.45 degrees north latitude and 72.40 degrees west longitude."

The old geographic standbys of latitude and longitude, often used to determine absolute location, can be confusing and difficult to work with. Many atlas publishers try to make it easier. The National Geographic Society's *Atlas of the World*, uses an alphanumeric grid with letters down the sides and numbers across the top and bottom of each map to help readers locate places. To locate Ulan Bator, Mongolia, for example, you would first look in the index of the atlas and see that it's on page 83, grid square D14. (The grid square system is similar to playing Bingo.)

Learning to Live with Latitude and Longitude

Lines of latitude and longitude form an imaginary grid on the surface of the earth that allows you to specify the exact location of any given place. Latitude indicates a location's north or south position on a map or globe; longitude indicates its east or west position. Both are stated in degrees and subdivisions of degrees. (Like a circle, a globe contains 360 degrees. Each degree is divided into 60 minutes of arc; each minute, into 60 seconds of arc.) When you specify a location using this grid system, latitude is given first and then longitude. Using both latitude and longitude, you can pinpoint any location on the earth.

Lines of latitude, more precisely called parallels, run east-west; lines of longitude, more precisely called meridians, run north-south. If you follow a meridian, you eventually run into the North or South Pole. If you follow a parallel, you head east or west and travel parallel to the equator.

Here's an overview of latitude basics:

- Latitude measures position north or south of the equator.

- The equator is 0 degrees latitude.

- The North Pole is 90 degrees north latitude.

- The South Pole is 90 degrees south latitude.

- Latitude ranges from 0 degrees to 90 degrees north or south.

- Always state direction (north or south) when you're specifying latitude.

Geographically Speaking

Longitude is not only essential in determining location, it is central to the world's method of keeping time. Because it takes the earth 24 hours to rotate once and the earth is a globe with 360 degrees of longitude, it's easy to figure out that it takes the earth an hour to rotate 15 degrees. By international agreement in 1884, the prime meridian, or 0 degrees longitude (though the French obtained and clung to the Paris Meridian for another 30 years), was declared to be the line running through the Royal Observatory at Greenwich, England. Starting at the prime meridian, 24 standard meridians spaced at 15-degree intervals were each designated as the center line of a one-hour time zone. For each time zone traveled east of Greenwich, the clock advances an hour. For each zone west of Greenwich, it loses an hour. When it's noon at Greenwich, it's midnight on the other side of the world at 180 degrees longitude. This line, where one day turns into the next, was designated the International Date Line. Although still roughly equivalent to the 180-degree meridian, the Date Line twists and turns to avoid splitting countries into different days. Although the time standard adopted in 1884, Greenwich Mean Time (GMT), has been superseded by the more precise Universal Coordinated Time (UCT), the principle is still the same.

And here's an overview of longitude basics:

- Longitude measures position east or west of the prime meridian, a north-south line that runs through Greenwich, England.

- The prime meridian is 0 degrees longitude.

◆ Longitude ranges from 0 degrees to 180 degrees east or west. Unlike latitude lines, longitude lines are not parallel because they come together at the poles.

◆ Always state direction (east or west) when you're specifying longitude.

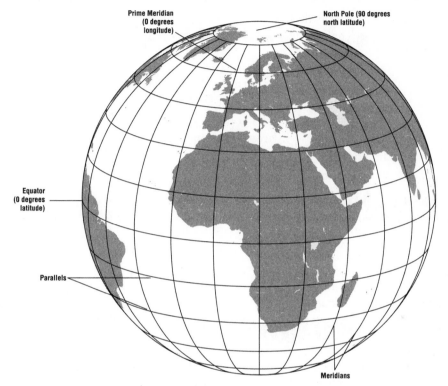

Lines of latitude and longitude.

Sizing Up Scale and Distance

Maps by necessity are reductions of the surfaces they represent; otherwise, in most cases, they'd be too huge to be of any use. So maps generally are drawn to a reduced scale. As you've already seen, scale is the ratio between a distance on a map and the corresponding distance on the surface of the earth. Perhaps a cartographer will use a scale of ½ inch on the map to represent 100 miles on the actual earth's surface, for example.

What Are Those Numbers at the Bottom of the Map?

A map's scale can be expressed as a ratio (1:24,000) or as a representative fraction ($\frac{1}{24,000}$). The first number in the ratio (the numerator in the fraction) represents a map distance; the second number (the denominator in the fraction) represents a ground distance. As long as the units you use are the same for both the map and the ground distance, you can plug in any unit of measurement (inches, feet, centimeters, or pencil lengths, for example).

Bellying Up to the Bar Scale

On some maps, an aid called a bar scale (sometimes called a graphic scale) allows you to use a ruler or piece of paper to measure between two features on the map and then compare that measurement to the bar scale at the bottom of the page to determine distance on the earth's surface.

The Rise and Fall of Elevation and Relief

Because the earth's surface is far from smooth, it is difficult to represent it accurately on a map. The earth is spherical (actually an oblate spheroid) and its surface irregular—a cartographer's nightmare.

Two terms are often confused in describing the ups and downs of the earth's surface. Elevation refers to the height (or depth) of the earth's surface above or below a baseline (usually sea level). Relief refers to the variation in elevation from one place to another. An area of extreme relief is an area with dramatic fluctuations in elevation.

Cartographers indicate elevation and relief on maps in many ways. Usually, shading and hatching (small parallel lines) are used to depict major mountain ranges or river valleys, while different colored tints indicate different elevation zones. Sometimes the actual height of a hill or mountain is printed next to its top (a spot elevation).

Perhaps the most accurate—and difficult to understand—method of indicating elevation and relief is with contour lines. A contour line on a map represents an imaginary, usually curvy line on the ground that follows a single elevation. This means that every point on the line is at the same elevation (an isoline or line of equal value).

A topographic map.

Section of 1:24000 scale, New Britain Quadrangle, Connecticut-Hartford Co., 7.5 Minute Series (topographic) United States Geological Survey, Reston, VA 22092, in cooperation with the Connecticut Highway Department.

Setting Off in the Right Direction

Direction is key to orienting yourself in the world and describing the location of one place in relation to another. The most common way of expressing direction is still north, south, east, and west. These are the cardinal (meaning "most important") points of the compass. Most maps have a compass rose indicating not only the four cardinal directions, but also four inter-cardinal directions (northeast, southeast, southwest, and northwest), and as many as twenty-four further subdivisions.

Terra-Trivia

If you're standing at the North Pole, the only direction in which you can travel is south.

North refers to the direction you follow to get to the North Pole. No matter where you are on Earth, if you set off toward the North Pole, you are heading due north. At a right angle to this north-south line lie east and west. On the face of a clock, if north is at 12 o'clock, then east is at 3 o'clock, south is at 6 o'clock, and west is at 9 o'clock.

Magnetic Attraction

We frequently use a magnetic compass to find north. This method has just one minor problem: in most places on Earth, the magnetic compass needle points to a different north than the geographic North Pole. A magnetic compass points to the earth's magnetic north pole. The North Pole through which the earth's rotational axis passes (where meridians converge and where Santa Claus lives) is a different place from the magnetic north pole.

In many cases, the difference between the true north of the North Pole and magnetic north (called magnetic declination) is not significant. It is mostly a concern for navigation and aviation while rarely causing confusion in day-to-day life. If it is a problem, one would use a special declination diagram. Because magnetic north moves around slightly from year to year, these diagrams are important in industry, defense, and long-distance travel.

Technical Directional

Azimuth is a more precise way to express direction than by using the points of a compass rose. Declination diagrams on topographic maps generally show directions as azimuths. Again, you need to visualize a clock face, but one that's divided into 360 degrees, as shown in the following figure.

An azimuth is measured clockwise from a north baseline (0 degrees or 12 o'clock on the clock face) to 360 degrees. Three o'clock, or due east, is a 90-degree azimuth; six o'clock, or due south, is a 180-degree azimuth.

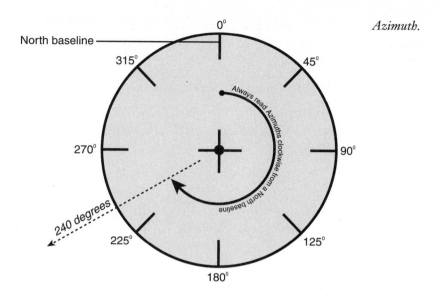

Azimuth.

New Directions in Mapping

The latest technology makes it possible to be even more precise in direction. The global positioning system (GPS) is a worldwide radio-navigation system that uses 24 satellites and their ground stations to calculate positions on the earth accurate to within yards now, to within inches in the not-too-distant future. Originally developed for the military, GPS has revolutionized marine navigation and is being used today to navigate and track vehicles, locate missing people, guide tractors, manage emergency responses, and increase the accuracy and versatility of maps.

Another technology that is transforming modern mapmaking is geographic information systems (GIS), a computer software/hardware system that links maps to databases and facilitates the manipulation, analysis, and display of spatial data. Think of it as an extremely accurate "super" map that can show relationships, change instantly to new parameters, and even predict how human-made and natural changes will affect a specific place.

The Least You Need to Know

- No map can represent the earth's surface without distorting size, shape, distance, or direction.

- Symbols are used to represent landscape features on a map; a symbol key (legend) explains their meaning.

◆ Relative location is useful for locating one position with respect to another. Absolute location identifies the position of a place by means of a grid system, such as latitude and longitude.

◆ Elevation refers to a feature's height above or below sea level. Relief is the variation in elevation.

◆ The cardinal directions north, south, east, and west correspond to the directions of meridians and parallels.

Chapter 6

Here We Go! A World of Many Regions

In This Chapter

- Identifying hemispheres
- Counting the continents
- Learning about regions

One way to get a handle on a confusing world is to divide it into parts called regions or subregions (also known as areas, territories, realms, ranges, fields, and domains). Although the terms are not important, you should understand each region's similarities and differences. This chapter identifies the major regions of the world and discusses the shared characteristics that qualify them as regions.

The Hemispheres: Half a World Away

If you choose the equator as the earth's dividing line (midway between the North Pole and South Pole), you can effectively divide the earth into two halves. A hemisphere is simply half a sphere. The northern half of the earth is called the Northern Hemisphere; the southern half, the Southern Hemisphere.

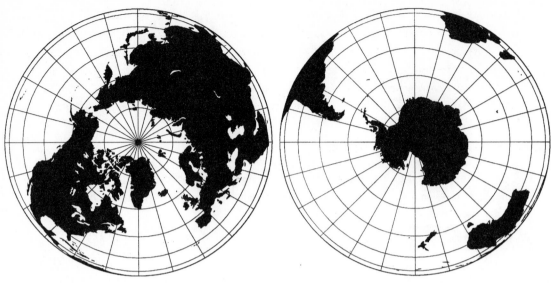

Northern Hemisphere.   *Southern Hemisphere.*

The earth can also be divided into an eastern half and a western half. But this is a more arbitrary division because any meridian can be used to divide the earth into two equal halves (the equator, on the other hand, is the only line of latitude that divides the globe equally). The usual line of demarcation between Eastern and Western hemispheres is either 0 degrees longitude (through the United Kingdom) and 180 degrees longitude (through the Pacific), or 20 degrees west longitude (through Iceland) and 160 degrees east longitude (through the Pacific). The advantage of the latter divider is that it leaves all of Europe in the Eastern Hemisphere and just the Americas in the Western Hemisphere.

The Seven Continents—Or Maybe It's Five? Or Four?

Although different geographers have proposed various criteria for defining a *continent*, a continent is generally considered to be a huge landmass surrounded entirely by water, a definition that works well for only about three of the usually listed seven continents.

The traditional seven continents are, of course, North America, South America, Europe, Asia, Africa, Australia, and Antarctica. Some experts argue, however, that North and South America comprise only one continent because they're connected by the Middle American land bridge and that Europe and Asia comprise one continent because they're a single landmass with no intervening body of water separating them. If you accept these arguments, the world has only five continents.

Or maybe just four. Even Africa doesn't fit exactly this book's definition of a continent. In the northeast, Africa is separated from Asia by the Gulf of Aden and the Red Sea. The Red Sea tapers into the Gulf of Aqaba and the Gulf of Suez. Only the thin, manmade Suez Canal connecting the Gulf of Suez with the Mediterranean Sea severs Africa from continental Asia. Some geographers go so far as to consider Europe, Asia, and Africa a single supercontinent sometimes (rarely) referred to as Eurafrasia.

The Europe/Asia/Eurasia Dilemma

Most of the boundary between Europe and Asia is not water but dry land. The following map shows the traditional dividing line between Europe and Asia.

The Europe-Asia boundary.

GeoRecord

The largest continent in land area is Asia, with 17.3 million square miles. It represents about 30 percent of the earth's total land surface, followed in size by Africa, North America, South America, Antarctica, Europe, and Australia.

The line between Europe and Asia runs along or through a combination of mountains, rivers, seas, and straits: the Ural Mountains, the Ural River, the Caspian Sea, the Caucasus Mountains, the Black Sea, the Bosporus Strait, the Sea of Marmara, and the Dardanelles Strait. Because the distinction between Europe and Asia is difficult to justify using any definition of the word "continent," many geographers simply refer to the continent of Eurasia.

Australia: Island or Continent?

Greenland is relatively huge (more than 840,000 square miles), but not quite huge enough for continent status. To be considered a continent, a landmass must be at least as large as Australia (nearly 3 million square miles). If it's any smaller than Australia, it's just an island.

Continents are tidy ways for geographers to divide the world into large chunks based on physical characteristics, but they usually are simply too big and too diverse to be studied as one unit, and they're not necessarily cohesive regions. To really begin to get a handle on the world, geographers break the world up into smaller chunks called regions.

Okay, So What's a Region?

Regions are hard to define exactly, but when you're identifying a region, you look for some degree of sameness. What physical, cultural, economic, or historical attributes make one area distinct from another? Geographic proximity is a factor, as political boundaries often are. Depending on the weight you assign to any particular factor, definitions of regions can and do vary.

Although a region may coincide with an entire continent, that's not usually the case. Regions usually encompass a much smaller area than an entire continent. Remember that the definition of a continent, as you read earlier in this chapter, has nothing to do with culture, history, economics, climate, or vegetation—all elements that factor strongly in the determination of a region.

One especially important consideration about a region, or a country in a region, is its level of development. Distinguishing between the developed or "peds," which used to be referred to as First World, and the developing or "pings," sometimes called Third World, requires a close look at economic factors, but also includes demographics, quality of life, and the social infrastructure. In general, North America, Western Europe, and Japan head the countries considered "peds," while most of Africa, South Asia, and South America are considered to be developing.

Geographically Speaking

Don't confuse the division of the world into continents with the Continental Divide (or Great Divide). This term refers to the elevated "backbone" of a continent that divides rivers flowing in one direction from those flowing in the opposite direction. In North America, the Continental Divide runs north-south roughly along the Rocky Mountain chain. Rivers to the east of the divide flow into the Gulf of Mexico and ultimately drain into the Atlantic Ocean; rivers to the west of the divide drain into the Pacific.

The Idiot's Approach

Rarely do any two geography books agree precisely on the breakdown of our world into regions. Nonetheless, the regions outlined here roughly parallel regional breakdowns you might see elsewhere.

For the sake of clarity, this book does not split up countries between regions. Country lines are used as regional limits, though many countries are quite culturally and physically diverse. It's also not uncommon for two countries bordering one another to fall into different regions.

The larger the region you attempt to describe, the more general your description must be. So the goal in this book has been to divide the earth into regions small enough and similar enough to allow you to make sense of their geography and understand what makes each a cohesive region.

On the following map, the earth is divided into 20 world regions:

A. Northern America	K. sub-Saharan Africa
B. Middle America	L. Middle East
C. South America	M. Central Asia
D. British Isles	N. Southern Asia
E. Western Europe	O. Eastern Asia
F. Northern Europe	P. Japan

G. Southern Europe

H. Eastern Europe

I. Russia

J. Saharan Africa

Q. Southeast Asia

R. Australia and New Zealand

S. the Pacific

T. the Poles

The size of the regions varies greatly: some are as small as the British Isles, others are as large as Russia. The populations of different regions also vary widely: Eastern Asia and Southern Asia have more than a billion people each; at the North and South poles, only a few scientists are hunkered down in Quonset huts.

Similarly, the physical attributes of each region are quite diverse, from the tropical rain forests of Southeast Asia to the icy mountains and fjords of Northern Europe to the desert expanses of Saharan Africa. By discussing each region in detail, this book gradually paints a portrait of the entire earth.

The 20 regions of the world.

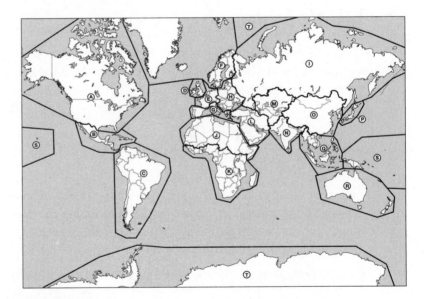

The Least You Need to Know

♦ A hemisphere is simply one half of the earth.

♦ Geographers traditionally have recognized seven continents: North America, South America, Europe, Asia, Africa, Australia, and Antarctica.

♦ Regions are areas of the earth that share a variety of geographic characteristics that make them distinct.

Part 2

A Regional Look: The Developed World

This part of the book examines the world's developed regions. The fact that you're reading this book means that you're probably from one of those regions, so at least some of the places described in Part 2 will be very familiar to you. Because these places are also the world's primary economic players, they're typically not far from the headlines.

Distinguishing between the "developed" and "developing" regions of the world isn't always easy. The point to remember is that a region's development level is neither fixed forever nor is it a value judgment.

Many "developing" countries are forging ahead with new technology and seizing a rapidly growing share of the world market (think China, for instance), while other older, "developed" nations may actually be in decline (think Japan, with its declining birth rate, lack of raw materials, and some rapidly advancing neighbors).

"Y'know—we could 'see North America' a lot faster on the Internet."

Chapter 7

Northern America: Land of Opportunity

In This Chapter

- ◆ Recognizing that the New World isn't so new
- ◆ Focusing on physical features
- ◆ Poring over population
- ◆ Examining ancestry, religion, and language
- ◆ Exploring the ongoing repercussions of September 11, 2001

The continent of Northern America is a place of immense variety, as reflected in its peoples, environments, and geographic features. This vast continent consists of Canada; the United States; Mexico; the countries of Central America, which are a land bridge to South America; the huge island of Greenland; and the islands of the Caribbean. Many geographers further divide the region into North America (composed of Canada, Greenland, and the United States) and Middle America (which includes Central America, Mexico, and the Caribbean areas).

The New World That Isn't

The Americas—North, Middle, and South—are often referred to as the New World, but this is incorrect, because the American continents were new only to the Europeans who stumbled upon them on their way to Asia. There were indigenous peoples in the Americas long before the coming of the Europeans, but this is another example of ethnocentrism, where one people claims superiority over others.

Northern America.

Two Countries, One Region

The United States and Canada share many cultural and economic features as well as the world's largest trading relationship and longest international open border (a political border that doesn't inhibit crossing—although as of January 23, 2007, passports are required for air travel between the U.S. and Canada). They share a lot of history, too. Both began as British New World possessions. Canada retains its ties to the British crown; the United States severed its connection in 1776. Both nations are highly industrialized and their populations enjoy very high standards of living. Because of their geographic proximity, intertwined economies, and cultural similarities, the United States and Canada constitute a natural and cohesive geographic region.

Canada: A Look at the Provinces and Territories of the Great White North

The Great White North, as Canada is popularly called, is a vast country with relatively few people. (Its entire population is less than that of California.) It is divided into ten provinces and three territories. The four small eastern provinces bordering the Atlantic Ocean are the Maritime Provinces: New Brunswick, Prince Edward Island (the smallest province), Nova Scotia, and Newfoundland and Labrador (which before 2001 was known as just Newfoundland). Moving west, you can see Canada's two largest provinces (in land area and in population): Quebec and Ontario. Farther west are the provinces known as the Prairie Provinces: Manitoba, Saskatchewan, and Alberta. All the way west, bordering the Pacific Ocean, is the tenth province, British Columbia.

Canada's territories lie to the north of the provinces and are large in land area and short on people. To the far northwest is the Yukon, which shares a long border with Alaska. East of the Yukon are the huge Northwest Territories and the newest territory, Nunavut, which was carved out of the Northwest Territories in 1999 as a homeland for the indigenous Inuit. (Nunavut is an Inuit word meaning "our land.")

The United States: 48 Plus 2 Equals More

Of the 50 United States, 48 form a single, uninterrupted landmass. These are the so-called "lower 48," also known as the 48 contiguous or conterminous states. (Contiguous means "touching or being in contact with each other"; conterminous means "sharing a boundary.") The remaining two states, however, are more far-flung. Alaska is separated from the lower 48 by Canada; Hawaii, by the Pacific Ocean. Geographers call countries like the United States that are not contiguous or conterminous fragmented countries.

> **Geographically Speaking**
>
> The United States consists of 50 states; a federal district (Washington, D.C.); an associated commonwealth (Puerto Rico); and a number of dependent areas, including the U.S. Virgin Islands in the Caribbean and American Samoa, Guam, Midway Islands, the Northern Mariana Islands, Wake Island, and other islands and atolls in the Pacific.
>
> Some of these areas are tiny; others are much more substantial. With a population of more than 3.5 million, Puerto Rico is larger than some states. American Samoa, Guam, the Virgin Islands, and the District of Columbia are represented in the House of Representatives by a delegate; Puerto Rico is represented by a resident commissioner. Although these representatives cannot vote on the floor of the House, they can vote on legislation in the committees to which they've been assigned.

Purple Mountains' Majesty and Fruited Plains

Extensive plains and great mountain chains dominate the Northern American landscape. The mountain chains, shown on the following map, all run roughly north-south, which can present an obstacle to anyone traveling east-west.

In Eastern North America, the dominant mountain chain is the Appalachian. Although these old, eroded mountains no longer reach great heights, they still stretch from Alabama to the Maritime Provinces in Canada. In the west, the higher Rocky Mountains and the Pacific ranges (the Cascades, Sierra Nevadas, and Pacific Coastal Range) run northward through the United States and Canada.

Along the Atlantic Ocean and the Gulf of Mexico are the coastal plains on which the largest concentrations of the region's people live. Stretching between the great mountain chains of the east and west are the vast central plains that extend northward past the Great Plains of the Midwest to the icy tundra of the far north. The central plains of Northern America are one of the world's most important breadbaskets, producing vast yields of grain year in and year out for domestic consumption and export.

The Mighty Miss

The Mississippi is the largest river in all of Northern America. Feeding the Mississippi River are its two largest tributaries (feeders), the Missouri and Ohio rivers. The tributaries of the Mississippi reach to the Rocky Mountains in the west, the Appalachian Mountains in the east, and just below the Canadian border in the north. They all drain south to the Mississippi's great delta and the Gulf of Mexico.

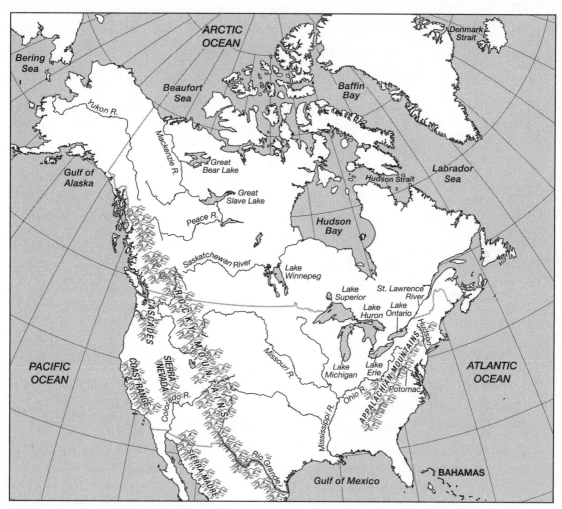

Northern America—physical features.

In addition to the Mississippi, noteworthy rivers in Northern America include the Columbia, Colorado (creator of the Grand Canyon), Mackenzie-Peace, Rio Grande, Saint Lawrence, and Yukon. Many smaller, well-known rivers lace the eastern part of the United States, including the Connecticut, Delaware, Hudson, James, Potomac, Savannah, and Susquehanna.

Water, Water, Everywhere

Within the landmass of North America are several of the world's largest lakes. The three oceans in this region are the Pacific to the west, the Atlantic to the east, and the icy Arctic to the north. The Bering Sea is notable because it and the narrow Bering Strait were all that separated the world's two greatest superpowers, the United States and the Soviet Union, during the cold war. This region's primary gulf, the Gulf of Mexico, is formed by the Caribbean Sea and the Atlantic Ocean. The gulf borders five states in the Southern United States as well as Mexico, for which it is named.

> **GeoRecord**
>
> The Mississippi–Missouri river (the combination of the Mississippi River and its largest feeder, the Missouri River) is the fourth-longest river in the world. At 3,740 miles long, it's exceeded only by the Nile, the Amazon, and the Chang Jiang (Yangtze) rivers.

Northern America also has 8 of the world's 12 largest freshwater lakes, including the 5 Great Lakes.

Weathering the Climate

Northern America stretches from near the tropics in the south to well within the Arctic Circle in the north, and these latitudes define the region's varied climate. The region is surrounded on three sides by oceans, which restrain climate extremes along coastal areas. Northern America's location in the Northern Hemisphere results in June-September summers and December-March winters.

Northern Canada and Alaska have cold polar climates. Summers are typically short and cool; winters are long and cold (not the ideal destinations for sunbathing). Most of the vast Northern American landmass experiences a range of humid climates, colder in the north and warmer in the south, with year-round precipitation. Southern Florida has a humid tropical climate, with typically dry winters. The Southwestern United States and the south-central plains of Canada have a dry climate, with arid conditions prevailing in the far southwest.

From Tundra to Mangrove

Northern America's vegetation generally parallels its climatic areas (as happens worldwide). Tundra (treeless Arctic plains) in the extreme northern reaches of the region gives way to pine forests as you move south. As you move farther south, the pine forests yield to deciduous forests of broadleaf trees that seasonally lose their foliage, while in the southeast pine forests again predominate.

Tallgrass prairies cover most of the drier central areas. The prairies, which once supported vast herds of bison and the Plains Indian culture that depended on them, are now primarily used for ranching. In Northern America's arid southwest, desert vegetation with cacti, shrubs, sage, and chaparral growth is the norm.

Another type of vegetation, unique to Northern America, is found in the Pacific Northwest. There, in old-growth temperate rain forests, are the largest trees on Earth. Redwoods, Douglas firs, Sitka spruces, and giant sequoias grow to be more than 300 feet tall in this area.

Black Gold

The United States has larger coal reserves than any other country on Earth. In Northern America, coal is abundant, accessible, and relatively cheap, but the environmental cost of extracting and burning coal—the "dirtiest" of the fossil fuels—is high. Despite large oil and gas deposits in the South Central and Southwestern United States and in Northern Alaska and Alberta, Canada, the United States is the world's largest petroleum importer and not coincidentally the largest producer of air pollutants and greenhouse gases.

Eye on the Environment

With petroleum reserves declining in both Canada and the United States, the U.S. economy is increasingly dependent on oil imports from some of the world's most volatile regions, particularly the Middle East. In an effort to lessen that dependence, the administration of President George W. Bush has embarked on an ambitious campaign to increase domestic oil production. A focus of that campaign is ANWR, the Arctic National Wildlife Refuge, an ecologically critical and pristine area on Alaska's northern coastal plain, which the administration proposes to open to oil drilling. Environmentalists argue that the ANWR reserves would hardly put a dent in U.S. oil needs and that its benefits would be far outweighed by the potential cost—to the area's indigenous peoples, wildlife, and ecology—of a major oil spill or other environmental disaster.

Mother Nature's Dark Side

Although Northern America is blessed with natural resources, it also has its share of natural disasters.

Terra-Trivia

The two countries of North America are both among the world's largest in total land area (Russia is largest). Canada is the world's second largest country, with 3,849,674 square miles of land area. The United States is the world's fourth largest country, with 3,618,770 square miles of land area.

Because the west coast of North America lies along the Pacific Ring of Fire, it is subject to volcanic eruptions and frequent earthquakes.

Droughts ravage the southwest; devastating tornadoes frequently plague the central plains states; the Mississippi River and its tributaries are subject to severe flooding; and the Gulf of Mexico and eastern coast of the United States are vulnerable to tropical storms and hurricanes. You might try Montana, where heavy snows and occasional forest fires are all you have to worry about.

Profiling the People

For hundreds of years, people from all across the globe have made their way to North America. They brought with them their ethnic characteristics, languages, customs, and religions. Because America is indeed a melting pot, any generalizations about its culture must necessarily be broad. After years of "melting," North American culture is unique.

Where the People Are

Population distribution in Canada is easy to visualize: think south. Almost all of Canada's relatively small population resides in the southernmost part of the country in a small east-west strip next to the United States. More than half of Canada's population lives in the area called Main Street, which runs from the north shore of Lake Erie northeast along the Saint Lawrence River.

In the United States, the largest population cluster is the northeastern megalopolis (basically Boston through New York City to the Baltimore/Washington, D.C. area). A second major population cluster extends west from the megalopolis and encompasses the manufacturing areas around the southern fringes of Lakes Ontario, Erie, and Michigan. Two more major population clusters are located in Florida and Southern California. Except for urban areas such as Denver and Salt Lake City, the interior of the United States remains relatively sparsely populated and is becoming even more so in some places. Rural areas of the Great Plains states, from Texas to Montana and the Dakotas, have been losing population for decades, a trend driven in part by the shift from small family farming to large-scale corporate agriculture.

The Urban Landscape

North America is a region of many large and well-known cities. Let's start with Canada:

◆ **Toronto, Ontario:** Largest city, financial center.

◆ **Montreal, Quebec:** Second largest city, the business and cultural center of French Canada.

◆ **Ottawa, Ontario:** Capital, seat of parliamentary government.

◆ **Vancouver, B.C.:** Canada's link to trade with the Pacific Rim.

◆ **Other important Canadian cities:** Quebec, Quebec; Winnipeg, Manitoba; Calgary, Alberta; Edmonton, Alberta.

At one time all major U.S. cities were located east of the Mississippi, most of them on the Atlantic Seaboard. The following list reflects a population shift to the west:

◆ **New York, New York:** The largest city in the Northern American region (Mexico City is the largest city in the continent of North America), New York is the financial and cultural capital of the United States.

◆ **Washington, D.C.:** The nation's capital.

◆ **Los Angeles, California:** Entertainment capital, pop-culture center.

◆ **Chicago, Illinois:** Home of some of the world's tallest buildings (the Sears Tower is North America's tallest building, at 1,454 feet).

◆ **San Francisco, California:** Tourist center, hub for high-level technology.

◆ **Other important U.S. cities:** Boston, Massachusetts; Philadelphia, Pennsylvania; Atlanta, Georgia; St. Louis, Missouri; Houston, Texas; Dallas, Texas; Detroit, Michigan; Denver, Colorado; Seattle, Washington; New Orleans, Louisiana; Phoenix, Arizona; and Miami, Florida.

Roots of the People

In both the United States and Canada, a large percentage of the population is of European ancestry. But both countries have also been magnets for immigrants from around the world, and today both have other large ethnic minorities. Canada has a substantial Asian population, concentrated mainly in the provinces of Ontario and

British Columbia. In Northern Canada, indigenous Indian and Eskimo ancestry is common. The United States has a large African American minority concentrated primarily in urban areas and throughout the south. As of 2003, however, Hispanics, or people of Latin American heritage, outnumbered African Americans to become the largest minority group in the United States. Hispanic communities are located in the cities and suburbs of the Northeast and Midwest, in south Florida, and in the states that border Mexico.

Indigenous Peoples

The native populations of the United States and Canada have shown remarkable resiliency in the face of overwhelming economic and educational disadvantages. Despite predictions at the beginning of the twentieth century that Native Americans would melt into the majority population and disappear, Indians have revived their tribal governments and cultural traditions and are asserting their legal rights. In 1999, Canada took a dramatic step toward reconciliation with its indigenous people by creating the new Inuit territory of Nunavut.

In the United States, the indigenous population of the Great Plains is on the rise, even though the general population is declining. Native Americans are leaving the cities in growing numbers and repopulating their ancestral homelands while some tribal groups have found economic growth by sponsoring gambling facilities on tribal lands.

Finding the Faith

Christianity is the dominant religion throughout much of Northern America. In the southeast and central portions of the region, Protestant faiths are most prevalent. Roman Catholicism is the dominant faith in the northeastern and southwestern sections of the region. Mormonism is common in the west central part of the United States (in and around Utah).

Judaism is practiced primarily in urban areas across Northern America. On reservation lands and in the far north, traditional Native American religions are being practiced with renewed vigor. There are also sizable Muslim populations throughout the region, as well as many followers of Eastern faiths.

Speaking of Language

Although the United States does not have an official language, English is the de facto official language and is spoken by the vast majority of its citizens. Although areas with growing Spanish-speaking populations are becoming increasingly bilingual, English is the norm for most of the United States.

Canada's language story is more complicated because it has two official languages: English and French. Francophones (French speakers) are concentrated primarily in the province of Quebec, where they constitute about 85 percent of the population. Quebec's struggle to preserve its French language and culture and keep it from being overwhelmed by English has fueled strong nationalist sentiment, and the province has on several occasions in the past seriously considered breaking away from Canada and forming a separate nation. Since the last referendum on separation was narrowly rejected in 1995, separatist feeling in Quebec is apparently on the decline.

In the News: September 11

With two mighty oceans on either side of them, the United States and Canada have long felt secure from the threat of foreign attack. That precious sense of security was shattered, perhaps forever, by the al-Qaeda terrorist assaults on New York City and Washington, D.C., on September 11, 2001. For the first time, the rage and violence of "foreign" religious strife swept onto American shores, and the region has not been the same since. With the threat of another terrorist attack still in the air, even that long-open border between two old friends and neighbors is not quite as open and free as it once was.

The Least You Need to Know

- The region of Northern America contains only two countries: Canada and the United States.

- Canada is politically divided into 10 provinces and 3 territories; the United States has 50 states, plus a number of outlying possessions and dependencies.

- Mountain chains in North America run north-south. The largest are the Appalachians and the Rockies.

◆ The Mississippi River is the largest and most important river in North America. The major water bodies surrounding North America are the Pacific, Atlantic, and Arctic oceans and the Gulf of Mexico.

◆ Canada is officially bilingual; the United States has no official language.

Chapter 8

The British Isles

In This Chapter

- A brief look at the rich history of the British Isles
- Distinguishing between Great Britain, the United Kingdom, and England
- Why Ireland is known as the "green land"
- Why everything revolves around London
- Whether the United Kingdom will give up its pounds for euros

The British Isles are located on the western fringe of Europe. Although their story parallels and meshes with much of Europe's history, their island status caused them to develop quite distinctly. Protected by water from every angle, Britain has withstood for nine centuries. The islands also dominated the surrounding seas for hundreds of years and achieved a power and greatness well out of proportion to their size.

Once Upon a Time

The British Isles have also had a disproportionate influence on world literature and science and played a leading role in the development of parliamentary democracy. British concepts of individual liberties and the legal system to ensure them are reflected in the constitutions and governments of many countries around the world. Until recently, this small grouping of islands ruled more than a quarter of the earth. Although the empire is gone, the region remains an industrial power. In the early twenty-first century, the British Isles are still a major global player.

The earliest inhabitants of these green, wet islands were the Celts, an Indo-European people. They were invaded and partially conquered by the Romans, and then absorbed wave after wave of Germanic tribes, including the Angles, Saxons, Vikings, and Normans. Along the way, the English language developed as a happy mixture of Celtic, Germanic dialects, Latin, and French.

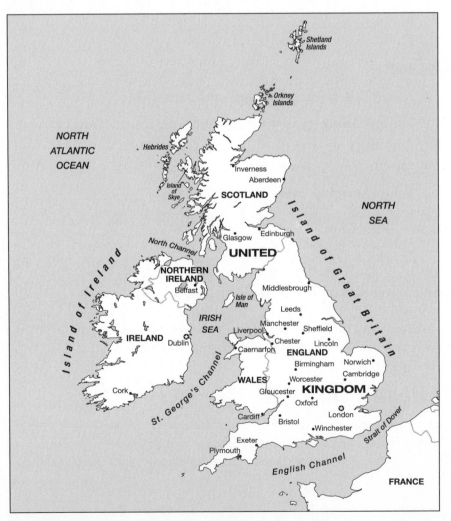

The British Isles.

The Empire

The British were always a seafaring people. They traveled the globe, trading and looting, settling and domineering. England came to dominate the British Isles and began to conquer the world's oceans after defeating the Spanish Armada in 1588.

Despite some setbacks, like losing their 13 American colonies in 1781, the British maintained a huge empire which included Canada, many Caribbean islands, India, Australia, and vast areas of Africa.

During the long reign of Queen Victoria, from 1837 to 1901, the United Kingdom reached the height of its power and influence. The British Empire extended around the globe and controlled one fourth of the world's land and one fifth of its people. The phrase "the sun never sets on the British Empire" was literally true.

With the passing of time, many conquered regions sought independence from this vast empire.

With the exception of six northeastern counties, which remained British, Ireland finally gained its independence in 1921, in the aftermath of World War I.

> ### Geographically Speaking
>
> The nineteenth century was a very low point for Ireland. Long oppressed by British colonial rule, a huge potato blight starting in 1846 brought about a massive famine. In the four years that the famine lasted, millions died or emigrated to the United States and other English-speaking countries.

World War II marked the beginning of the end of the rest of the British Empire. Independence came for most of the British Empire, which remained loosely aligned with the "Mother Country" through the British Commonwealth of Nations. Some countries, including Canada, Australia, and New Zealand, recognize the British monarch as their sovereign while independently functioning as democracies.

The Recent Past

The two world wars of the twentieth century seriously depleted Britain's economic strength. But since the end of World War II and the dissolution of the empire, Britain has reinvented itself as a modern industrial, high-tech European society. Its economy is still one of the world's biggest, and as one of five permanent members of the United Nations Security Council and a founding member of NATO, it is a major player in Europe and the world. In 1973, the United Kingdom and Ireland belatedly joined the European Union, then known as the European Community. Britain has always been ambivalent about its integration with Europe. Even now that it is a key member of the European Union, the United Kingdom, unlike Ireland, has held off joining the European Monetary Union, as this would entail giving up the mighty pound sterling in favor of the upstart pan-European currency, the euro.

The United Kingdom—It Doesn't Equal Great Britain

The region of the British Isles is composed of two large islands and a collection of smaller, peripheral islands. Although the entire region is known as the British Isles, only one island is named Britain. The two large islands are …

- **Great Britain:** The larger, more easterly island is sometimes called just Britain. It includes three constituent countries of the United Kingdom (England, Scotland, and Wales).

- **Ireland:** Although Ireland is the name of the smaller, more westerly island, it's also the name of a country that occupies most of the island, along with the six counties of the north, Ulster or Northern Ireland, which is a constituent country of the United Kingdom.

The two countries that occupy the region are:

- **United Kingdom:** Also called the United Kingdom of Great Britain and Northern Ireland, and is a political union of the four constituent countries of England, Scotland, Wales, and Northern Ireland. The United Kingdom occupies the entire island of Great Britain and the northeast portion of the island of Ireland.

- **Ireland:** The Republic of Ireland, or Eire, is the smaller of the two countries the make up the region. Ireland, independent since 1921, occupies all of the island of the same name, except the northeast corner which, as noted, is part of the United Kingdom.

At various times in history, each constituent country of the United Kingdom was a separate political entity and two have been independent kingdoms in their own right. Although the four are joined together as the United Kingdom of Great Britain and Northern Ireland, letting go of the rivalries and grievances of the past is not always so simple.

◆ **England:** The largest and historically the most dominant of the constituent countries that form the United Kingdom, England occupies the bulk of the southern and eastern portions of the island of Great Britain.

◆ **Scotland:** Like England, Scotland was once an independent kingdom. Scotland occupies the northernmost third of the island of Great Britain.

◆ **Wales:** Situated on the southwestern portion of the island of Great Britain, Wales never developed into a fully integrated kingdom as did Scotland and England. The Celtic principality was joined by force to England back in 1284, but has retained its distinctive language, culture, and sense of nationhood through the centuries.

◆ **Northern Ireland:** All of Ireland and Great Britain were united constitutionally in 1801 as the United Kingdom of Great Britain and Ireland. This became the United Kingdom of Great Britain and Northern Ireland in 1928 after the island of Ireland was divided into the independent Republic of Ireland and British Northern Ireland.

Although England, Scotland, Wales, and Northern Ireland are called constituent countries, none of them are separate, independent nations. In recent years, however, Scotland, Wales, and Northern Ireland have acquired a considerable amount of autonomy from the central government in London, especially since the establishment of a Scottish parliament and assemblies for Wales and Northern Ireland in 1999.

The United Kingdom still includes a number of holdovers from the empire. These include Bermuda, the British Virgin Islands, the Cayman Islands, the Falkland Islands, Gibraltar, Pitcairn Island (settled by mutineers from the ship *Bounty*), and Saint Helena.

> **Terra-Trivia**
>
> Each of these constituent countries contains smaller political subdivisions. England and Wales have counties; Scotland has regions; and Northern Ireland has districts. The country of Ireland also has counties.

Cool Britannia

Throughout the United Kingdom, smokestack economies have largely given way to a booming service economy, focused on banking, insurance, and business services. England, especially London and its surrounding counties, was a prime beneficiary of the tech explosion of the 1990s.

The United Kingdom has large reserves of coal, natural gas, and petroleum. Its agriculture is highly mechanized and relatively efficient, compared to that of Europe. Although the economic repercussions of the September 11, 2001, terrorist attacks in the United States and the bursting of the tech bubble in the early 1990s had a dampening effect on the UK economy, it remains one of the largest and strongest in Europe.

Eye on the Environment

The nineteenth century saw massive industrial growth in the British Isles, making them world leaders in the production of steel and other heavy manufactured goods. This economic success came with a high price, however. The so-called "Black Towns" in the British industrial areas were literally covered in thick coal residue from the many blast furnaces. This caused tremendous air pollution, pollution of rivers and land, and many respiratory illnesses and deaths.

The Bonnie Land of Scots and Lochs

Though sharing a country with the English, the Scots have remained distinct. Helped by their highland location and tight clan system (which is the basis of family life), they have maintained a rugged character and an independent approach to life. The Scots maintain a strong sense of nationalism.

The coal and iron once mined in Scotland have been depleted, and manufacturing in general continues to decline. Scotland has rebounded, however, with research and development, electronics, and North Sea oil and gas. Tourism is also a mainstay.

The Slate and Coal of Wales

The Welsh traditionally eked out a living by mining coal and slate or working the machinery of the industrial revolution. Aside from strip-mine scars on the landscape, there are few reminders left of the principality's recent industrial past. In the 1980s, overseas investment by multinational firms in the electronics, automotive, and chemical industries began to transform the Welsh economy. The services sector is also playing an increasingly important role here. Still, Wales lags behind the other regions of the United Kingdom in terms of economic development and prosperity.

Both Wales and Scotland have seen a resurgence in nationalism in recent years. Road signs bear both English and Celtic names, and the traditional languages are increasingly being taught in school. The new Welsh national assembly and Scottish parliament are both an expression and a fulfillment of these national aspirations.

The Troubles of Northern Ireland

Northern Ireland was settled primarily by Scottish and English immigrants who crossed the Irish Sea to occupy one sixth of the territory of the island of Ireland. Unlike the almost totally Catholic Republic of Ireland (Eire), only a third of Northern Ireland is Catholic; the rest of the population is primarily Protestant.

There is an ongoing turmoil in Northern Ireland as the Catholics, often using terrorism and violence through the Irish Republican Army (IRA), and the Protestants, also employing terror tactics through their various paramilitary groups, seek dominance in the area; this ongoing civil strife is referred to as "The Troubles," and has caused untold misery and economic deprivation for the people of the area.

Although there have been some promising events, the two groups remain antagonistic. With a large British military force helping maintain order, the Protestants wish to continue their close ties with England, while the Catholic groups hope for a closer relationship with Eire.

Ireland: The Green Land

Ireland was always very agricultural and was one of the least industrial areas in the British Isles.

Today, however, agriculture has been eclipsed by industry and Ireland has transformed itself into a modern, trade-dependent economy with one of Europe's fastest growth rates. Tourism is another important sector of the economy, as Ireland's green hills and lush countryside draw travelers from around the world.

Natural Diversity

The British Isles offer a great deal in the way of natural diversity. During the ice age, much of the region was enveloped by massive ice sheets, which left their mark on the northern landscape. Because the region consists of islands, the surrounding waters play a significant role in modifying its climate.

Landforms: From Scottish Highlands to Peat Bogs

The Northwestern Uplands, or Highlands, of Scotland are as high and spectacular as those in Norway, but are too rocky and wet for all except the hardiest settlers.

Terra-Trivia

Separating Scotland from England is the famous 75-mile-long Hadrian's Wall. Built in England by the Romans in 200 C.E., it was designed to keep northern barbarians at bay. It symbolically separates the Scots and English even now.

The Highlands continue south from Scotland, turning into the Pennine Mountains in north central England and the Cambrian Mountains in Wales. The north country is also sprinkled with lakes that form the magnificent Lake District, celebrated by the romantic lake poets William Wordsworth and Samuel Taylor Coleridge.

Glacial remnants endure in the jagged, fjord-like terrain of Northern Scotland. Bogs and moors abound as glaciers from the Grampian Mountains poke their way past lochs and through heather coastward, where the fishing is rich and the people more numerous. Aside from the Highlands of Scotland and Wales, most of the rest of the island of Great Britain is lowland, aptly called "downs." The lowlands are low (not surprisingly) and arable (easily cultivated) and have adequate rainfall. Marking the westernmost edge of Europe, the Emerald Isle of Ireland is lush and green and rimmed with low hills spaced around the fertile central plain. Half of Ireland is made up of meadows; the other half is filled with lakes, mountains, and the peat bogs that cover one seventh of Ireland.

The British Isles include many smaller islands, most of which you can see on the maps in this chapter. To Scotland's northeast lie the Orkney Islands, replete with race tides, skerries (small, rocky islands), and rocky reefs; only 20 of the 70 are inhabited. Farther to the northeast are the Shetland Islands.

On Scotland's northwest coast are the more than 500 Hebrides Islands. In the Irish Sea between Great Britain and Ireland is the Isle of Man. The Channel Islands are in the English Channel off the coast of France; and off Land's End, in extreme southwest England, are the tiny Isles of Scilly.

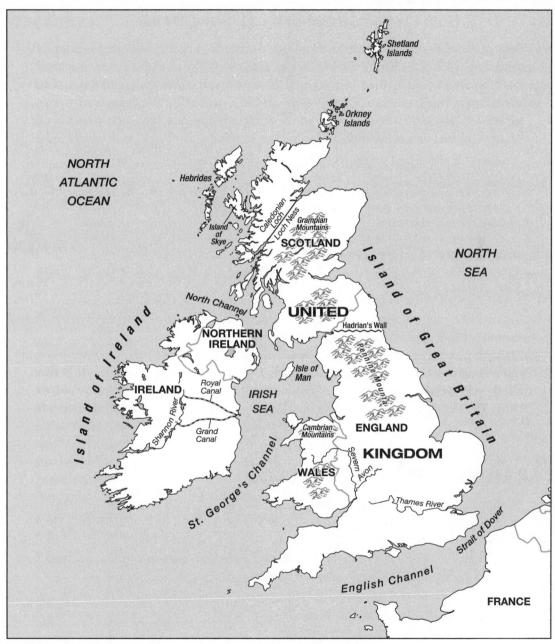

British Isles—physical features.

Waterways: From the Gulf Stream to the English Channel

The primary bodies of water surrounding the islands are the Atlantic Ocean to the west and the North Sea in the northeast. Separating the British Isles from continental Europe are the English Channel and the Strait of Dover. Ireland is separated from Great Britain by the North Channel in the north, the Irish Sea in the middle, and Saint Georges Channel in the south. Great Britain and Ireland are also home to many well-known rivers, such as the Severn, Thames, and Avon in Great Britain and the Shannon in Ireland. Canals abound, including the 60-mile-long Caledonian Loch that slices across Scotland. England is laced with hundreds of miles of canals that connect virtually every corner. Ireland features two famous canals, the Royal Canal and the Grand Canal.

A Balmy Climate at 55 Degrees North?

England is known for its fog and rainy skies, and Ireland is even rainier. What's surprising is that the region isn't known for its heavy snows. Great Britain stretches from about 50 degrees to 59 degrees north latitude. Carried over to North America, the latitude would approximate that of Hudson Bay, with a climate that's ideal for polar bears; however, Britain is much warmer because of the Gulf Stream, a vast, warming ocean current. Originating in the tropical waters of the Gulf of Mexico, it becomes the North Atlantic current that warms the British Isles. This huge flow of warm water tempers the region's climate and helps to make it milder and wetter than its latitude would otherwise dictate.

The Cities: London and Beyond

London is the dynamic hub of the British Isles. It's the capital of the United Kingdom and one of the world's largest and most cosmopolitan cities. Straddling the Thames River, which gives it access to the North Sea, London is also a large and bustling port. It is a global center of finance, publishing, communications, and retailing, and one of the world's great tourist attractions and cultural centers.

The region has more cities than just London, however. Here are some of the more notable:

England

- ◆ **Liverpool–Manchester:** This conurbation (network of cities) was a leader during the industrial revolution. Manchester was known for its textiles, and

Liverpool served as a major port. Manchester, one of England's largest cities, has become a major air transportation hub and commercial center; Liverpool remains a great shipping port and industrial center.

◆ **Birmingham:** England's second largest city was badly damaged by World War II bombing and has not yet fully erased the scars of its post-industrial decline. But Birmingham, now England's biggest convention center, is sprucing up and becoming a greener and more enjoyable city than in the recent past.

Scotland

◆ **Glasgow:** This old industrial city benefits from oil and gas exploration, industrial research, and electronics.

◆ **Edinburgh:** The capital and cultural center of Scotland.

Northern Ireland

◆ **Belfast:** Despite the sectarian violence that has plagued Belfast, the Northern Ireland capital is a surprisingly happening place, with a lively economy and vibrant nightlife.

Wales

◆ **Cardiff:** Largely rebuilt after World War II bombing, the Welsh capital on the Severn River is rapidly developing into one of Europe's finest maritime cities, famous for its parks, choirs, and ancient Cardiff Castle.

Republic of Ireland

◆ **Dublin:** The once dowdy Irish capital has in recent years become the Celtic "Silicon Valley" and the hub of Ireland's recent economic boom.

The People and Their Languages

One legacy of the British Empire is the large number of ethnic minorities who live in the United Kingdom, mostly Chinese, West Indian, and South Asian. Although church attendance is down dramatically here as in most of Western Europe, most people are at least nominally Protestant. The largest denomination is Anglicanism, followed by Presbyterianism and Methodism. There is a large Catholic minority, and sizable numbers of Muslims, Hindus, Sikhs, and Jews. The Republic of Ireland is overwhelmingly Roman Catholic. Both islands are densely populated, except for the northern reaches of Great Britain.

Most people in the United Kingdom and Ireland speak English, a Germanic-based language. Three remnant Celtic tongues are still spoken on the islands. Wales has the complex Welsh language. In Northwestern Scotland, highlanders still speak the Gaelic language. Irish is spoken throughout Ireland, especially in remote clusters on the western coast.

In the News: The Euro Future

The integration of the United Kingdom into the European Union was late in coming and has not been altogether smooth. A sizable percentage of the British population was against joining the European Union in the first place back in 1973, and as of 2003 a majority is against joining the European Monetary Union and adapting the euro. One argument against the euro is that the UK economy is stronger than the Eurozone economies and would be dragged down by monetary union. Proponents of the euro argue that the Eurozone wields too much economic clout and is too important to British trade for the United Kingdom to remain out of it for much longer.

The Least You Need to Know

- The British Isles are made up of many small islands and two large ones, Great Britain and Ireland. The region consists of two countries, the United Kingdom and the Republic of Ireland.

- Despite the high latitude of the British Isles, the influence of the Gulf Stream makes its climate fairly mild.

- London is by far the largest city in the British Isles and a major world financial and cultural capital.

- The people of the United Kingdom are largely Protestant; Ireland's population is overwhelmingly Roman Catholic.

- English is the dominant language in the region, along with remnants of ancient Celtic tongues.

Western Europe: The Heart of the Continent

In This Chapter

◆ The mixed blessing of this region's location

◆ The heart of Europe

◆ The natural side of Western Europe

◆ The region's spectacular cities

The Western European region has been the mother of great philosophers, scholars, writers, artists, traders, and explorers. Though it's home to only a fraction of the world's population, it's a dominant player in trade, manufacturing, finance, art, fashion, and technology.

Small but complex, Western Europe is a collage of separate nations characterized by different economies, politics, and physical features. Although old animosities, religious differences, and language barriers have not completely gone away, Western Europe is at the core of the increasingly cohesive and expanding European Union.

Blood and Glory: Western Europe's Turbulent Past

Western Europe's location on the continent has proved to be both an asset and a curse. The region has served as a crossroads for millennia and has thrived on the trade and culture that such a strategic position fosters. The downside is that, as a crossroads, Western Europe has for centuries also served as the battleground of the Western world.

Western Europe.

Countries in the Crossroads of Europe

Western Europe can be grouped into three primary geographic subregions:

- ◆ **France and Germany:** Core countries that form the largest subregion, in terms of land area and population

- ◆ **Switzerland, Austria, and micro Liechtenstein:** Landlocked Alpine countries

- ◆ **Netherlands, Belgium, and tiny Luxembourg:** The "Low Countries" that comprise the smallest subregion

The Biggies: France and Germany

The largest of Western Europe's countries (it's not quite as large as Texas), France covers more than 200,000 square miles of varied terrain. It is the westernmost country in the region, and with nearly 60 million people, it is the second most populous. The French countryside is laced with vast networks of rivers and canals that unite the country and connect it with the rest of the continent.

France is Western Europe's leading agricultural producer. Grains from the north (France is the only European country able to export surplus wheat) are the dominant crop. Farms in the south benefit from long growing seasons to produce olives and oranges. And of course, France is the source of the world's finest wines.

France also rules supreme in the design, production, and retailing of luxury goods and fashion. French brand names, such as Chanel, Dior, Givenchy, Saint Laurent, Vuitton, and many others, are international icons of chic. Germany, the home of bratwurst and beer, Volkswagen automobiles, and the speed-limit-less autobahn, is called the "locomotive" of Europe. The continent's most populous nation and largest economy, Germany has one of the world's greatest concentrations of industrial centers and factories and dominates the European coal, iron, and steel industries. Every type of machine—military equipment, oil pipelines, and even roller coasters—is made by German industry, which in turn is dependent on coal and oil.

The industrial heart of Germany, and of Europe, is the Ruhr Valley in the northwestern part of the country. From such industrial centers as Essen, Duisburg, and Dortmund, manufactured goods move easily via Rhine waterways to Rotterdam in the Netherlands and out to sea. But as in other parts of the developed world, the industrial pulse of the Ruhr Valley is slowing as the economic focus of Germany's economy shifts to commerce and services.

Geographically Speaking

After separating East and West Germany for 28 years, the Berlin Wall, the hated symbol of Communist oppression in the east, finally came down in 1989. West Germany and East Germany were united politically once again in 1990, but the unification of their two very different economies has been slow and difficult. The cost of bringing the infrastructure and economy of the former East Germany up to Western standards has been far greater than anticipated and has weakened Germany's economic performance in recent years.

Germany relies heavily on a low grade of coal to fuel its industry, and emissions from its coal-burning utilities contribute to air pollution. Environmental problems are especially acute in the former East Germany, where acid rain and other industrial pollutants are to blame for what is known in German as waldsterben, or forest death.

The Alpine States

In the southeast of the Western European region lie the Alpine states. This subregion consists of three countries: Switzerland, Austria, and the microstate of Liechtenstein. As the term Alpine states implies, these countries are dominated by Western Europe's vast mountain chain, the Alps.

Terra-Trivia

The Swiss are known for their fierce independence and strict neutrality. They stayed out of both world wars in the last century, they have not joined the European Union, and while they have long participated in United Nations organizations, the country did not officially join the UN until 2002.

Switzerland, like its Austrian neighbor, is famous for its Alpine beauty and picturesque qualities. This small landlocked mountain country is one of Europe's most prosperous, with a highly skilled work force, a strong currency, and a secretive banking system that attracts foreign investment.

Having few energy resources, the Swiss have harnessed hydropower and nuclear power to fuel the renowned Swiss precision industry. Machine instruments, watches, and diesel engines make up the bulk of Swiss exports.

Unlike Switzerland, Austria was at the center of both twentieth-century world wars, and despite being on the losing side both times, it has emerged as a modern, prosperous nation with a thriving, mainly service, economy. Like Switzerland, Austria uses "white coal" (hydropower), of which it is a net exporter, to fire its machine and chemical industries. Long a leader in forest management, Austria has been careful in using its Alpine forests. Because the forests protect against erosion and avalanches, they have been harvested selectively to provide fuel and lumber—no clear-cutting here.

The microscopic principality of Liechtenstein is set high in the Alps astride the Rhine River and between Switzerland and Austria. Small though it may be, Liechtenstein has a big economy, focused on industry and financial services. Gorgeous mountains and fairy-tale castles make Liechtenstein a popular tourist destination.

The Low Countries

The third subregion of Western Europe is composed of three small, low-lying countries located in the north central part of the Western European region. The Netherlands, Belgium, and Luxembourg are known collectively as the Low Countries or the Benelux countries (from the combination of the first two or three letters of *Bel*gium, the *Net*herlands, and *Lux*embourg).

The Netherlands, which literally means "low lands," is also known as Holland, although strictly speaking that's the name of two of the country's provinces (North and South Holland). The people of the Netherlands, or Holland, are called the Dutch in English. The Netherlands are traditionally associated with windmills, tulips, and wooden shoes, but what really defines the country is water. With two fifths of the Netherlands lying below sea level, much of Dutch history has been about flood prevention. Although water has played a destructive role in Dutch history, it has also made the country's fortune. Situated at the mouths of three major rivers—the Rhine, the Maas, and the Scheldt—the Netherlands is one of the world's most important commercial and transportation hubs. Its main port, Rotterdam, is Europe's largest. The Netherlands is also a major producer and exporter of food and flowers, and of course, mountains of colorful Edam and Gouda cheese.

Belgium is situated south of the Netherlands and is in the middle of the Low Countries in terms of land area and population. The country is divided between French-speaking Walloons in the south and Dutch-speaking Flemings in the more industrialized north. Although conflicts between the two groups once threatened to tear the country apart, increased regional autonomy has helped ease tensions considerably. Belgium today is the bureaucratic heart of Europe, with NATO and the European Union both headquartered in its capital, Brussels.

The last of the Low Countries in this list is the pseudo-low country of Luxembourg. At about the size of Rhode Island, Luxembourg is just a notch above being a microstate. Despite its reputation as an outdoor paradise filled with lush forests and rushing streams, Luxembourg has a very modern, high-growth economy. Although steel was once king in Luxembourg, its economy has diversified, with the focus increasingly on financial services. Multilingualism and stability make Luxembourg attractive for foreign business.

The expanding European Union.

The U.S. of E? The European Union Today

The countries surveyed in this chapter form the core of what is today known as the European Union (EU), formerly known as the European Community (EC) and before that, the European Economic Community (EEC). Germany and France are its largest and most populous countries and are two of the four trillion-dollar economies in the European Union. (The other two are the United Kingdom and Italy.)

The idea for a European Federation was first proposed in France in 1950, but the foundation for it had been established in 1948, when the three Benelux nations joined together in a customs union. The original six members of what would become known as the European Economic Community (the Benelux countries, plus Italy, France, and Germany) became nine in 1973 when the United Kingdom, Ireland, and Denmark joined the European Community. As of 2003, the European Union included 15 member states (Greece, Spain, Portugal, Austria, Finland, and Sweden joined in subsequent expansions) and took in another 10 in 2004 and 2 more as of January 1, 2007, for a total of 27 members. After the 2004 expansion, the European Union included for the first time former Soviet-bloc countries of Eastern Europe.

Despite rifts and occasional setbacks, the European Union has brought about a prosperous, unified, border-free, economic, and increasingly political entity stretching from the Atlantic to the borders of the old Soviet Union. In January 2002, 11 of the 15 member states cemented their economic union by exchanging their historic currencies—francs, marks, liras, pesetas, escudos, shillings, drachmas, guilders, punts, and crowns—for the single European currency, the euro.

Western Europe's Natural Side

Pushing into the Atlantic Ocean on the west and the North Sea on the north, much of Western Europe has a maritime feel. The moderating influence of the Atlantic Gulf Stream tempers the plains of the northwest. Interior and mountain areas face colder climes. Protected by mountains from the cold of the north, the French Riviera (the Mediterranean shore of France) enjoys sunny temperatures with only a little rain.

Landforms: From Polders to Alps

Contrasting with the lowlands of the Netherlands is the most striking of Europe's physical features: the mighty Alps. This rugged chain is a 680-mile-long collection of snowy peaks and valleys filled with glaciers and dotted with lakes. The Alps form an arc that stretches from France's Mediterranean coast through Switzerland and Germany to Austria's Danube.

Two great mountain chains separate Western Europe from Southern Europe. The Alps run between Western Europe and Italy, while the Pyrenees Mountains form a natural boundary between Southern France and Spain. The Pyrenees are the home of the Basques, a people whose language is related to no other and whose homeland straddles the French and Spanish sides of the western Pyrenees. The Basque struggle for a separate state has periodically involved violence and terror attacks.

Despite the size of these huge and magnificent mountain chains, they have not completely barricaded nor isolated one region from another. Because of natural passes, valleys, and rivers, north-south trade, communications, and invasions have been a fact of life for millennia in this part of the world.

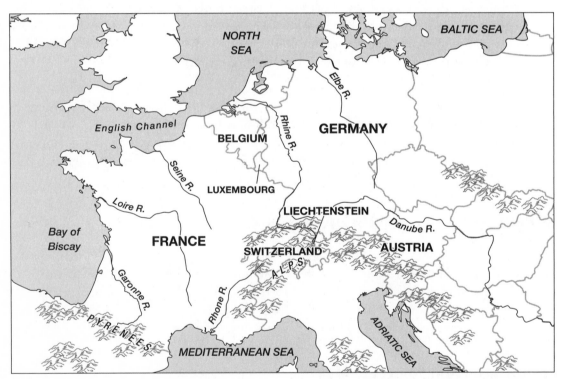

Physical features—Western Europe.

Creating Land from Water

The Netherlands is cradled around what was once a shallow inland sea called the Zuider Zee. At the mouth of this sea, the Dutch built a huge earthen dam, or dike, that cut off the Zuider Zee from the North Sea. The dike created a lake from the former Zuider Zee that the Dutch renamed the IJsselmeer. Over the years, the Dutch have drained portions of this shallow lakebed to form reclaimed lands called polders.

The poldering process that has enabled the Dutch to create land out of what was once the sea involves the construction of massive dikes around areas to be reclaimed. After the dikes have been constructed, the brackish water is pumped out and the area's famous canals are then used to drain the land. (Windmills once had the job of constantly pumping out the ever-seeping water. Today diesel pumps are used and the windmills are primarily scenic elements.)

North Sea storms have breached the dikes in the past, and the loss of life and property has been catastrophic. After a devastating flood in 1953, the Netherlands began the construction of a cutting-edge system of dikes, dams, and surge barriers on the mouth of the eastern Scheldt River. The Delta Works, as the project is called, is one of the engineering wonders of the world.

Waterways: From the Danube to the Rhine

Snowmelt from the Alps gives rise to three of Europe's key rivers: the Rhine, the Rhône, and the Danube, as shown on the preceding map. Other notable rivers are the Seine, the Loire, and the Garonne, which together with the Rhône form the basis of France's great waterway system. The flat, level plains of Northern France and the Low Countries have made it easy to join the area's major rivers—the Rhine, Meuse, Seine, and Marne—through a complicated network of bridges, rail links, and canals.

Western Europe's most significant river is the majestic Rhine, which starts in the Swiss Alps, flows through Western Germany (where it forms part of the France-Germany border), and empties into the North Sea at the great delta in the Netherlands. The Rhine is the major lifeline of the Western European region, carrying goods and people from the industrial heartland out to sea.

A Place of Wondrous Cities

Many of the world's most spectacular and storied cities are located in the Western Europe region. There are too many to adequately discuss, but here are a few highlights.

France

◆ **Paris:** The seat of government and the primary expression of national pride and culture. Rich in historical and cultural treasures, including the great Gothic cathedral of Notre Dame and the magnificent Louvre Museum (home of da Vinci's Mona Lisa), Paris is one of the world's most beautiful cities.

◆ **Lyons:** With its prime location on the confluence of the Rhône and Saône rivers, Lyons is second only to Paris in population and economic clout. It is a historical center of silk and textile manufacturing and site of major annual trade fairs.

◆ **Bordeaux:** This port on the Garonne River is a major cultural center and the heart of one of the most important wine-producing regions in the world.

Germany

◆ **Berlin:** This vibrant city of 3 million people is once again the capital of a reunified Germany. From 1961 to 1989, Berlin was divided by the most notorious expression of the Cold War, the Berlin Wall.

◆ **Hamburg:** Germany's second largest city, Hamburg is a major port at the mouth of the Elbe River and one of the world's greenest cities.

◆ **Munich:** The main city of Southern Germany and Bavaria, Munich is the site of the world's most famous annual beer festival, its raucous Oktoberfest.

Switzerland

◆ **Zurich:** Switzerland's largest city is also a global center of banking and business.

◆ **Bern:** The capital and seat of Swiss government.

◆ **Geneva:** The home of the Red Cross and the League of Nations.

Austria

◆ **Vienna:** Famous for its musical heritage and rich desserts, Vienna is now the capital of small but prosperous Austria, but was once the center of the mighty Hapsburg Empire.

The Netherlands

◆ **Amsterdam:** The art-rich capital of the Netherlands is known as the "Venice of the North" for its many canals and magnificent town houses. Amsterdam shares capital honors with The Hague, the seat of the Dutch government and site of the World Court.

◆ **Rotterdam:** Almost completely destroyed during World War II, Rotterdam today is one of the world's largest and busiest ports. Rotterdam is known as the "gateway to Europe" because of its strategic location on the mouth of the Rhine.

Belgium

◆ **Brussels:** Beyond its magnificent medieval central square, the Belgian capital is a booming modern city that serves as headquarters for NATO and the European Union.

The People

Although it's not a large region in area, Western Europe does have a very large population. Especially heavy concentrations of people live in Northern France, the Low Countries, the Rhine River valley, and along the Rhône in Southern France. Western Europe is highly developed and industrialized, and most of its people live in cities.

The Netherlands, with more than 1,200 Dutch squashed into each square mile, is the most densely populated country in Europe (with the exception of the microstates of Malta and Monaco); Belgium is a close second. Because Europe's birthrate is relatively low and life expectancy is high, however, you might hear mutterings about a "demographic winter" whose projections suggest a high old-to-young ratio. As this ratio becomes reality, Western Europe faces a situation of having fewer workers and higher costs for social programs.

In Germany, the declining population trend is counterbalanced somewhat by the continuing influx of *Gastarbeiter*, or guest workers, who have been filling jobs in Germany since the 1960s. Immigrant workers from Turkey, Greece, Spain, and more recently Russia and the former Yugoslavia have also kept the country's growth rate on the positive side (barely). Without this immigration, Germany's population would actually be shrinking.

Like Britain, France had an overseas empire as well. Immigrants from former French colonies in Southeast Asia and North Africa, especially Algeria, have settled in the cities and suburbs of France, where ethnic prejudices and rivalries have led to profound cultural conflicts.

Protestants and Catholics

Most of the Western European region is nominally Christian, although church attendance is in decline here as in the United Kingdom. The north is predominantly Protestant; the south is mainly Catholic. The Dutch, Swiss, and Germans are fairly equally divided between Protestant and Catholic. The French, Austrians, Belgians, and Luxembourgians are primarily Catholic. Although the Jewish population of Western Europe was decimated during the World War II holocaust, the world's third largest Jewish population (about 600,000) lives in France.

Languages: Germanic and Romance

Western Europe has two primary language subfamilies: Germanic and Romance. Both subfamilies belong to the Indo-European language family. French, the primary representative of the Romance subfamily, is spoken primarily in France. French is also spoken in Southern Belgium and Western Switzerland.

The Breton Peninsula in northwestern France is considered a refuge area. It has an isolated pocket of people who speak Breton, a language in the Welsh group of the Celtic subfamily. Celtic languages are relic languages. Once spoken over a wide area, they were overwhelmed by other languages and now survive only in refuge areas.

The Germanic languages (including Dutch, Flemish, and German) are dominant in the remaining countries in the north and east. Almost all Europeans are at least bilingual, with much of the population proficient in three or more languages. Switzerland has four official languages: Italian, French, German, and Romansch (an umbrella term for several Romance dialects spoken on both sides of the Swiss-Italian border). By the way, many Western Europeans have also mastered English.

The Least You Need to Know

- ◆ Western Europe has been the site of major wars throughout history.
- ◆ The European Union emerged from the devastation of World War II.
- ◆ The Alps are the dominant physical feature of the region. The Rhine is its principal river.
- ◆ The people of Western Europe are primarily urban, affluent, and Christian. They speak either Romance or Germanic Indo-European languages.

Chapter 10

Northern Europe: Land of the Midnight Sun

In This Chapter

◆ Europe's rugged northern fringe

◆ Legacy of the Vikings

◆ Great cities of the north

The region of Northern Europe stretches northward from the continental core of Europe through the Baltic Sea and the North Atlantic Ocean to the frozen Arctic Circle. Northern Europe is the northernmost group of countries on Earth. The sea has provided a livelihood, albeit a difficult one, but has distanced the people of the north from the rest of Europe. The result is a sense of neutrality and unity—and no small degree of isolation.

Democratic and literate, sturdy and industrious, Northern Europe's sparse population is homogeneous, mostly Protestant, and enjoys a high standard of living. An egalitarian outlook on life is typical, and the "cradle to grave" social support system is an accepted part of life.

Northern Raiders and Resisters

Beginning in the eighth century, Viking warriors from Northern Europe carried out raiding expeditions throughout a large part of the continent. But the Vikings were more than raiders: they explored and discovered as they traveled, and Viking settlements expanded into Greenland and Iceland, eventually making them the first Europeans to land in North America. In the seventeenth century, the Swedish kingdom was at the height of its power and extent, encompassing all of present-day Sweden, much of Finland, a large part of the Baltic region, and sections of Germany east and west of Denmark. Eventually, other nations dominated the open seas and the Scandinavians stayed closer to home.

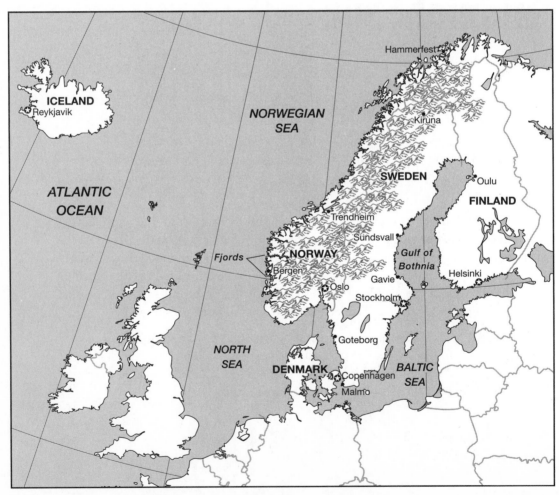

Northern Europe.

The Land of the North

Europe's northern tier includes five countries, which as a group are referred to as Norden. Scandinavia refers to the Scandinavian peninsula containing Norway and Sweden. Finland, Denmark, and far-off Iceland (900 miles away) complete the Nordic group. The five Nordic countries have a great deal in common: a shared heritage and culture, rugged landscapes (except for low-lying Denmark), closely related languages (except for Finnish), strong economies, and a long and close association with each other. The result is a very distinctive and cohesive geographic entity within Europe.

Norway, Sweden, and Finland all lie in the Northwestern Uplands, Europe's oldest mountain range, which runs all the way to Great Britain. The countries of Northern Europe benefit from the moderate temperatures generated by the Gulf Stream (a warm current) as it pours into the North Atlantic. Southern coastal Norway does not have a frigid climate; harbors remain open and free of ice year-round. The same cannot be said, of course, of the interior or mountain regions, where the weather, like the terrain, is much more severe.

Despite some relief from the Gulf Stream, the region still encroaches on the Arctic Circle. Winter days in this region are short, dark, and cold, with lots of wind and moisture. The area has poor, rocky soil and a short growing season. The far north is wild and mountainous, with much of it above the Arctic Circle. Most of this area is blanketed with vast stands of coniferous forests, called taiga, and with extensive tracts of birches.

The Fjords of Norway

Norway is famous for the magnificent fjords that define its jagged and indented western coastline. A fjord is a long, narrow inlet, or arm of the sea, created by glaciers and lined by steep cliffs. Some are as long as 100 miles. Fjords are fed by river waterfalls that cascade from the mountains to join the sea. As the fjords reach the sea, skerries (small, rocky islands) form a buffer between the Atlantic and the shore.

The thousands of rivers and streams from Norway's high mountain ranges produce more kilowatt hours of energy per person than anywhere else in the world. In addition to hydroelectricity, the country enjoys other resources in the form of oil and natural gas yielded by the North Sea. Half of Norway's exports are petroleum products, and a great deal of industry has developed around power sources, both hydro and petroleum.

The Swedish Core

On Sweden's west and north sides, rugged highlands mark the boundary with Norway. Coniferous forests grow in the northeast, and large lakes and low agricultural land are found in the south. The country is laced with rivers that originate in the highlands of the west and drain to the Gulf of Bothnia in the east. Although lacking Norway's oil and gas deposits, Sweden does possess substantial mineral wealth, primarily in the form of metal ores.

The Forests of Finland

Finland's terrain is flatter than that of its neighbors to the west, but poor soils and cold temperatures make agriculture nearly impossible. Mineral resources are few. Although this situation sounds dismal, Finland does have trees, and lots of them, which account for its large and thriving paper, pulp, and lumber industries.

Although nearly a third of the country lies above the Arctic Circle (its neighbor to the east is Russia), Finland, like its neighbors, benefits from the warming influence of the Gulf current. Despite its northern location, Finland never has permanent year-round snow cover or ground frost. The waters of the Baltic Sea and the Gulf of Bothnia also contribute to the relative mildness of Finland's maritime climate. Some 30,000 islands form micro-archipelagos (island clusters) along the Finnish coast.

Denmark

As the southernmost Nordic country, Denmark occupies yet another peninsula, Jutland. In the south, the Jutland peninsula shares a mere 42-mile border with Germany. The low flatlands of the Jutland peninsula strike a sharp contrast to the rugged highlands of the Scandinavian peninsula to the north.

Also notable are Denmark's 482 islands. They range from mite-size to medium-size to Greenland, the world's largest island. In the midsize category is Zealand, home to Copenhagen, the capital. Denmark also has the 21 Faeroe Islands off Scotland's north shore. Some of the Faeroes, meaning "sheep islands," have a language and flair of their own and even a separate set of laws. All make up the Kingdom of Denmark.

The country is more agriculturally blessed than its Nordic cousins. With productive soils, milder temperatures, level lands, and high-tech methods, Danish farmers are net exporters of food, with an emphasis on meat and dairy products. With farmland that covers three quarters of the countryside, Denmark is indeed the Nordic breadbasket.

Iceland's Heat from Below

Of the five Nordic countries, Iceland is the worst off in terms of quality terrain. Three quarters of the interior is filled with volcanoes, glaciers, lava formations, frozen lakes, and varied wastelands. Most of Iceland, in fact, is uninhabitable. All the island's settlements, and the few farms that make hay for the famous Icelandic ponies and hardy sheep, are coastal. Perched on the spreading mid-Atlantic ridge, a volcanic hot spot, Iceland treats its 100 volcanoes as an additional energy resource. Geothermal heat from volcanic sources and hydroelectric power from Iceland's rivers and waterfalls are used to warm apartment buildings as well as greenhouses for vegetable production.

Eye on the Environment

Although most of Iceland's electricity and heating needs are now met by hydroelectric power and geothermal sources, its cars, buses, and all-important fishing fleet are still powered by conventional fossil fuels, of which it has none. To free itself from its dependence on imported fuel, Iceland is at the forefront of what might be the coming hydrogen revolution.

Hydrogen is an efficient, clean-burning fuel that is found wherever there is water, and there's certainly a lot of water in and around Iceland. The problem with hydrogen (H) is that extracting it from water (H_2O) requires electricity and lots of it, and that usually means burning (dirty) fossil fuels to produce (clean) hydrogen. And that's where Iceland has an edge: its electricity is abundant, cheap, and clean. Using it to extract hydrogen to power vehicles and ships makes both environmental and economic sense. Iceland's first energy revolution involved hydroelectric power; its second involved geothermal heat. Its third will most likely involve hydrogen.

The Descendants of the Vikings

Considering Norden's location at the top of the world, you probably aren't surprised to hear that few people live there. The hardy souls who do reside there live on the coasts and as far south as possible. With just fewer than 9 million inhabitants, Sweden boasts the largest population in the region. Population densities are fairly low throughout Norden (Denmark has the highest), and the bulk of the people are urban dwellers.

The Lapps: Life on the Top of the World

Along the Arctic Circle across the top of Norway, Sweden, Finland, and part of Russia live the aboriginal people of Norden, the hardy, nomadic Lapps, or Sami, as they prefer to be called. The Sami traditionally have made their living hunting, fishing, and herding reindeer on the frozen, moss-covered, treeless arctic plains known as tundra (a Sami word). In recent years, however, increased mining and logging activities are putting pressure on traditional Sami reindeer-herding lands.

Terra-Trivia _____

With fewer than 280,000 residents, Iceland is one of the world's least populated countries (the population of the entire country is less than that of Birmingham, Alabama). Population density in Iceland is a paltry seven people per square mile. To put this number in perspective, Iceland has about 18,160 fewer people per square mile than Hong Kong and only 7 more people per square mile than the moon.

Almost Intelligible

Most of Norden shares a common linguistic heritage. Danish, Icelandic, Norwegian, and Swedish are all Germanic languages in the Indo-European language family, and all are more or less mutually intelligible. Finnish, however, is the Nordic oddball. It originated in central Asia and is part of a separate language family, the Uralic family, to which the Sami language, Estonian, and Hungarian also belong.

The Many Religions of Northern Europe?

Think Lutheran. Although the Nordic region today is largely secular, its main religion is Lutheranism.

Making Ends Meet on the Tundra

Northern Europe today has emerged as one of the continent's most prosperous regions. The economies in the north have thrived on fishing (especially Iceland), wood products (especially Finland), industry and manufacturing (especially Sweden), agriculture and food processing (especially Denmark), and petroleum revenues (especially Norway). The people of Northern Europe generally are well educated and have high incomes.

Geographically Speaking

Iceland shares a maritime economy with its Nordic neighbors. It sits amid the richest fishing grounds in existence and annually delivers more than 1.5 million tons of fish to the rest of the world. Because the fish industry provides 70 percent of Iceland's export earnings, it has no interest in having foreign fleets encroaching on its domain and takes great pains to make that information public knowledge. It has also resisted joining the European Union for that very reason.

The Cities of the North

Scandinavia has a number of towns and cities, but in each case the capital of these countries is the primate city (largest in size and characterizing the essence of each nation).

Norway

 ♦ **Oslo:** Norway's capital and center of commerce and culture. Nature-loving Oslo is one of the world's most heavily forested cities.

Sweden

 ♦ **Stockholm:** The capital of the most industrialized of the Nordic nations, Stockholm was built on islands where the Baltic meets Lake Malar, hence its renown as "the queen of the waters."

Finland

 ♦ **Helsinki:** Finland's capital and largest city, Helsinki, known as the "white city of the north," is famous for its striking modern architecture.

Denmark

 ♦ **Copenhagen:** The capital of Denmark and the most cosmopolitan of Nordic cities, Copenhagen is home to the world-famous amusement park Tivoli Gardens, which features symphonies and ballets, glitzy rides, and tranquil gardens.

Iceland

 ♦ **Reykjavik:** The Icelandic capital is a small but booming city, home to more than one in three Icelanders. One of Europe's newest tourist hotspots, Reykjavik is also the world's northernmost national capital.

In the News: Bridging the Ore Sund

The new Ore Sund Bridge between Copenhagen and the Swedish port of Malmö provides the first fixed link between Sweden and Denmark (and by extension the European heartland) since receding glaciers split them apart 7,000 years ago.

Opened to traffic in July 2000 after nearly 10 years of planning and construction, the Ore Sund Bridge is the world's longest single road and railway bridge.

The Least You Need to Know

♦ Five countries are in the Northern European, or Nordic, region: Norway, Sweden, Finland, Denmark, and Iceland.

♦ The Nordic countries are the most northern countries in the world.

♦ With the major exception of North Sea oil and gas, the region is relatively poor in fuel and mineral resources, but is highly developed and affluent.

♦ The region's relatively small population is located in the south and along the coasts.

Chapter 11

Southern Europe: Embracing the Mediterranean

In This Chapter

♦ The Greco-Roman heritage

♦ The Southern European quartet

♦ Glorious cities of art and history

♦ The peoples and cultures of Mediterranean Europe

The four countries of the Southern European region share intertwined pasts and common cultural roots. Greece is the fountainhead of Western civilization, and the imprint of ancient Rome is evident throughout Italy and the countries of the Iberian peninsula, Spain and Portugal. The Southern European quartet also shares a Mediterranean climate and vegetation, mountainous landscapes, and generally poor soil. All either occupy or share peninsulas.

There are exceptions, of course, to all this sharing. Portugal, Spain, and Italy are predominantly Roman Catholic and speak Romance languages. Greece is Orthodox and speaks a non-Romance language. Greece and Italy are fully in the Mediterranean Sea, whereas Spain also faces the Atlantic, and Portugal faces only the Atlantic. Although Italy is one of Europe's four trillion-dollar economies, Spain, Portugal, and Greece are among the least developed of the current European Union member states.

The Legacy of the Greeks and Romans

The ancient Greeks established an extensive trading network that spanned the Mediterranean. The achievements of the Greeks were later absorbed and expanded on by the Romans.

Southern Europe.

The area was regularly invaded by various peoples, including Muslims, Vandals, Visigoths, and Lombards.

Southern Europe has long been a leader in the arts, science, and exploration. The Renaissance began here, and sailors from this region were instrumental in starting the age of exploration and expansion of Europe to the Americas, Africa, and Asia.

The Countries of Southern Europe

As shown on the preceding map, Southern Europe consists of four countries and several microstates. In the west are Portugal and Spain; in the center lies the famous boot of Italy; and to the east is Greece, with its many islands.

Seafaring Portugal

Portugal is surrounded by Spain and the Atlantic Ocean. Occupying 35,550 square miles of the Iberian Peninsula, Portugal is Southern Europe's smallest and westernmost country. At one time, Portugal was the center of a vast overseas empire that at its peak was 23 times the size of the motherland. Its major colonies included Brazil in South America; Angola, Guinea-Bissau, and Mozambique in Africa; Goa in India; Macao on China's southeast coast; and East Timor in Southeast Asia. The last remnant of Portugal's empire was Macao, which it returned to China by agreement in 1999.

Income levels in Portugal are lower than in most other European Union countries. Although the Portuguese economy has expanded and diversified since it joined the European Union with Spain in 1985, agriculture is still a mainstay of its economy. Portugal's farmers raise sheep, cattle, and poultry, and grow typical Mediterranean crops: olives, fruit, grapes, and grain.

Terra-Trivia

Portugal is a world leader in the production of cork. Cork is extracted from the outer bark of the Mediterranean cork oak tree, which abounds in Portugal.

Spain—Land of Toros and Tradition

Spain is Southern Europe's largest country in land area and its second most populous (after Italy). It is more urbanized and industrialized than Portugal and enjoys a higher income level, although Spain, too, is in the lower economic tier of the European Union. As with the other countries in the region, agriculture figures prominently in the Spanish economy. Both Iberian nations are major wine producers: Portugal is famous for its Vinho Verde and port wines; Spain is known for its Riojas and sherries.

Tourists from Northern Europe and beyond flock to the cosmopolitan cities, Moorish-flavored towns, and sun-drenched beaches of Spain.

Italy: Europe's Cultural Magnet

Italy is the giant of the region. Although second to Spain in area, Italy has the largest population and the greatest population density in the Southern European region. It is also the Mediterranean's economic leader, with the bulk of the region's industry and its highest income levels. Its economic health, however, is clouded by the persistent disparity in income levels and standard of living between the prosperous north and the poorer, more agricultural south. Other hard-to-shake problems confronting Italy include government corruption and the influence of organized crime.

Greece and the Aegean

Greece is a land of archeological wonders and shimmering islands set in the blue Aegean. Given these two resources, it is not surprising that tourism accounts for 15 percent of the country's gross domestic product. A whopping 20 percent of its workforce is involved in agriculture, even though only a quarter of the land is arable, plots are small, and rainfall is unreliable. Nevertheless, Greek farmers produce tobacco, cotton, and wheat in addition to the typical Mediterranean mix of crops: olives, grapes, oranges, and vegetables. Although Greece has benefited enormously from membership in the European Union, its income levels still lag behind those of most other EU members. Greece, especially the Athens area, went through a major infrastructure overhaul as the country prepared to host the Summer Olympic Games in 2004, but has seen little improvement of its general economy from the successful Olympics.

Greece was part of the empire of the Ottoman Turks for centuries before it gained its independence in 1829. Bad feelings between Greece and its neighbor Turkey linger to this day, even though the two are members of the North Atlantic Treaty Organization (NATO) and Turkey seeks to join Greece in the European Union. Although confrontations between the two countries have diminished in recent years, a lingering bone of contention between Greece and Turkey is the island of Cyprus, which has been divided into Greek (Orthodox Christian) and Turkish (Muslim) sides since 1974. Directly and indirectly, many Greeks are tied to the sea. The population is clustered largely along the coasts, and even now many fish for their livelihood. Shipping is also an economic mainstay, and Greece boasts a substantial merchant marine fleet.

The Microstates

In addition to its quartet of large countries, the Southern European region includes a handful of tiny (but full-fledged) countries called microstates:

- **Monaco:** Located in the southeast corner of France, on the dazzling French Riviera right near Italy, this infinitesimally small constitutional monarchy is occupied almost entirely by the city of Monte Carlo. Monaco is famous for its casinos and its jet-set residents and visitors.

- **Andorra:** This miniscule microstate is sandwiched between Spain and France, where Catalan-speaking farmers tend to their goats deep in the Pyrenees.

- **Malta:** South of Sicily, this tiny island did not become a republic until 1974, when the British departed after almost 200 years of military presence.

- ◆ **San Marino:** Perched atop Italy's Mount Titano, San Marino is an independent republic in the middle of Italy.

- ◆ **Vatican City:** In the heart of Rome, this Lilliputian state, wrapped around Saint Peter's Basilica, is the home of the Pope and the spiritual center for more than 1 billion Roman Catholics worldwide.

The Lay of the Land: Peninsulas and Mountains

As you can see from the following map, Southern Europe can be likened to a series of peninsular fingers poking into the Atlantic, the Mediterranean, and the region's assortment of lesser seas. The largest of these peninsulas, the Iberian Peninsula, is located at the region's western edge. Occupied by Spain and Portugal, Iberia is bounded on the north by the Pyrenees Mountains, which separate the peninsula from France. To the south, only eight miles and the Strait of Gibraltar separate Iberia from Africa. The interior of the Iberian Peninsula consists largely of high, rugged plateaus.

To the east of Iberia is the boot-shaped peninsula of Italy. A mountain range also separates this country from Western Europe: This time, it's the Alps. Notable passes through the Alps include the Brenner from Austria, Maloja from Switzerland, and Little St. Bernard (yes, of canine fame) from France. Running the entire length of the "boot" is Italy's spine, the Apennine Mountains. To the south, on the island of Sicily, is the still-active volcano Mount Etna.

Greece is located on the southern tip of the Balkan Peninsula on Southern Europe's eastern edge. This is rugged, mountainous country. From the jagged southern end of the Dinaric Alps and the Balkan Mountains, the Pindus Mountains run south through the peninsula, dominating the Greek landscape.

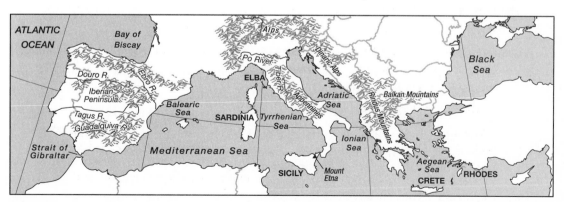

Southern Europe—physical features.

South of the Balkan Peninsula is yet another peninsula, the Peloponnesus, site of the famous Peloponnesian Wars between ancient Sparta and Athens. The isthmus of Corinth connects, or rather connected, the Peloponnesus to the mainland. Since the Corinth Canal was cut through the isthmus in the late nineteenth century, the Peloponnesus, technically speaking, has become an island.

Islands Everywhere

In addition to its three main peninsulas, Southern Europe abounds in islands. Greece wins the prize for the most islands, with more than 2,000 of them making up the Greek archipelago (string of islands). The largest of the Greek islands are Rhodes, Lesbos, and Crete.

Italy also has its share of islands, including the large islands of Sicily and Sardinia. Spain controls the Balearic Islands (Mallorca is the best known) on its Mediterranean East Coast and also owns the Canaries off the West Coast of Morocco. Portugal's islands include the Azores and the Madeiras (as in the wine), located to the west and southwest in the Atlantic Ocean.

Warm Waterways

The Mediterranean Sea dominates Southern Europe. Only Portugal has no coastline on this sea (its coastline is exclusively on the Atlantic Ocean). The Mediterranean is subdivided into several smaller seas.

Spain has coastlines on the Atlantic Ocean on the west and on the Mediterranean's Balearic Sea on the east. To the south is the Strait of Gibraltar, which connects the Atlantic and the Mediterranean. To the north of Spain lies an offshoot of the Atlantic Ocean, the Bay of Biscay.

As Italy juts southward into the Mediterranean, it edges the Adriatic Sea on its northeast and the Tyrrhenian Sea on its southwest. Between the sole of the Italian "boot" and Greece is the Ionian Sea. The bulk of the Greek islands are located in the Aegean Sea between mainland Greece and Turkey.

Southern Europe has several rivers of note, as shown on the map earlier in this chapter. Italy's major river is the vital Po, whose basin is home to the bulk of the country's agriculture, industry, and population. Of great historical significance is the Tiber River in central Italy, on which Rome was founded.

Spain also has several significant rivers. The Ebro rises in the Cantabrian Mountains in north central Spain and cuts deep gorges in Catalonia before emptying through a wide delta into the Mediterranean. Because northern Spain uses the Ebro (or Spanish Nile) for irrigation, the Ebro River basin has developed into a productive and populous area. The Guadalquivir River in the southwest passes through Córdoba before flowing to the Atlantic. The fertile lowlands of the Guadalquivir Valley are one of Spain's primary agricultural areas. Originating deep in the heart of Spain is the Tagus River, which passes through Toledo before bisecting Portugal and emptying into the Atlantic Ocean at Lisbon.

Climate: The Mediterranean Sun

Almost all of Southern Europe enjoys a Mediterranean climate. Hot and dry summers prevail, and the brown land lies dormant until the winter brings cooler and wetter days. Northern mountainous areas throughout the region can be cold and snowy. On the Spanish interior plains, arid is the word of the day, and temperatures are extreme. Endlessly flat and treeless, this interior region called secano (unirrigated) is by no means barren. Fields of corn, rows of gnarled olive trees, and vineyards crawl over the prairies of La Mancha. The common vegetation of the entire Southern European region is what the Italians call *macchia*, the French, *maquis,* and the Spanish, *matorral.* Similar to what Californians call *chaparral,* this dense, scrubby growth is able to survive arid conditions thanks to water-holding thorns and leaves. The maquis has replaced the native forests of the region, which were long ago burned or cut for lumber. Remnants of these once prolific forests, however, survive in Italy's Po River valley.

The Historic Cities

The cities of Southern Europe are treasure-houses of art and history.

Greece

- ◆ **Athens:** The glory of the Greek capital is its wealth of classical antiquities, beginning with the hilltop citadel and religious center, the ancient Acropolis. Around the Acropolis is a sprawling, congested, smog-filled city that can be exasperating. In recent years, however, this *primate city* has made strides in

def•i•ni•tion

Athens is a classic example of what geographers call a **primate city,** a country's largest and most important city historically, culturally, and economically. A primate city is often the capital of its country.

controlling its pollution and congestion. Piraeus, the port of Athens, handles most of the country's massive amount of shipping.

Italy

◆ **Rome:** The eternal city offers a mix of the ancient, medieval, Renaissance, and modern eras. Built on the famous seven hills, Rome embraces its ancient lifeline, the Tiber River. The city is Italy's capital and is home to spectacular piazzas, churches, palaces, and of course the vestiges of its imperial past.

◆ **Milan:** Unlike cities farther south where the twentieth century is less in evidence, Milan is a fast-paced commercial city and financial center, complete with skyscrapers and an extensive subway system. Although it is home to the gothic Duomo and da Vinci's Last Supper, it is one of the world's major capitals of fashion and contemporary design.

◆ **Venice:** Built on more than 100 islands in a lagoon off Italy's northern Adriatic coast, Venice is one of the world's most beautiful and art-filled cities. Today, however, the Queen of the Adriatic faces multiple crises. The increasingly flood-prone city is sinking at an alarming rate.

◆ **Florence:** The Renaissance and, in many respects, the modern age got their starts here. Florence was the home of Dante, Donatello, Ghiberti, Michelangelo, and Galileo, to name a few.

Spain

◆ **Madrid:** Situated on Spain's high central plateau, the national capital is a political and financial center and industrial hub.

◆ **Barcelona:** This is the wealthy, sophisticated capital of the Catalan region in Northeastern Spain. It is an important seaport and center of commerce.

Portugal

◆ **Lisbon:** From its harbor at the mouth of the Tagus River, Portugal's capital once ruled a trading empire that spanned the globe. Today, Lisbon is reinventing itself as a modern, vibrant European city, where soaring skyscrapers jostle with medieval churches and magnificent relics of Portugal's proud seafaring past.

The Peoples of the South

As might be expected in so mountainous a region, the people in this region have tended to cluster in river valleys and along the coast; this is especially true in Greece. More Southern Europeans derive their livelihoods from agriculture than is typical elsewhere in Europe; urbanization, not surprisingly, is also lower than the European norm.

Roman Catholic and Greek Orthodox

Although the Roman Catholic and Orthodox churches share common origins, the Great Schism of 1054 C.E. split them apart for good. Southern Europe is overwhelmingly Roman Catholic, with the notable exception of Greece, which is almost entirely Greek Orthodox.

Romance Languages—and a Few Others for Good Measure

Romance languages aren't the only tongues spoken in this region; in fact, not all the languages spoken there even belong to the Indo-European language family.

The Romance branch of the Indo-European family is dominant in three of the four countries in the region. In Portugal, virtually everyone speaks Portuguese. In Spain, most people speak either Castilian Spanish or Catalan. The latter is a distinct Romance language spoken by some 10 million people, primarily in Northeastern Spain (Valencia, Catalonia, the Balearic Islands, and parts of Aragon), as well as in Andorra and in parts of Southern France and Sardinia. In the far northwest portion of Spain, people of Celtic stock speak Galician, which is related to Portuguese. The Basque language, spoken in north central Spain, is an odd bird discussed on the following page.

Italy is also in the Romance camp. Although dialects and regional differences exist throughout the country (especially the Sicilian dialect to the south), Italian is the language of the land. The only exceptions are pockets of Illyrian languages and in the far north, where Romansch (another Romance tongue) and German are spoken.

Greece is the only major country in the region not to speak a Romance language. Greek is in the Hellenic branch of the Indo-European language family, and is spoken across almost all of Greece and its islands. It might not be a Romance language, but at least it's in the Indo-European family.

The same thing can't be said of Basque. Unique to Southern Europe, Basque is not a Romance language and is unlike any other language known today—in fact, it constitutes its own language family. Iberia's oldest native inhabitants, some 50,000 Basques live in a beautiful pocket of the Pyrenees in northern Spain. With their unique language and Asian nomadic stock, the Basque people tend to be taller and fairer than most Spaniards. Dissatisfaction with centralized control from Madrid has spawned a powerful separatist movement among the Basques that has often spilled over into violence and terrorism.

In the News: Illegal Immigration

The countries of Southern Europe have traditionally been net exporters of labor, sending emigrant workers to North America and Western Europe in search of better jobs and better lives. As living standards in the Southern European region rose in the latter part of the twentieth century, and especially after Greece and the Iberian countries joined the European Union, migration patterns reversed dramatically. No longer net exporters of workers, each country in Southern Europe is now viewed as a destination as well as a transit point to Western Europe by emigrants from Africa, Asia, Eastern Europe, and South America.

The nations of Europe find themselves in a tricky bind with regard to illegal immigration. On the one hand, their native populations are growing slowly if at all and are increasingly reluctant to take on low-wage jobs. As current demographic trends play themselves out, Europe will become even more dependent on foreign labor, legal or otherwise, than it is now. On the other hand, unorganized immigration brings with it a wide range of problems, from organized crime and drug trafficking to illegal arms smuggling and terrorism. In the post-9/11 world, secure borders are much more than just a legal formality, they can be matters of life and death.

The Least You Need to Know

- The Greeks and Romans left a lasting cultural imprint on Southern Europe.

- All four countries in the region (Portugal, Spain, Italy, and Greece) are located on peninsulas and have many islands. Greece has more than 2,000 of them.

- All the Southern European countries except Portugal border the Mediterranean Sea, and all share a Mediterranean climate.

- Higher living standards and membership in the European Union have made all four Southern European countries magnets for illegal immigration.

Japan: The Pacific Dragon

In This Chapter

- Tracing the sword of the samurai
- Touring Japan's islands, mountains, and plains
- Meeting the island-nation's people
- The economic giant and its future

The Japanese name for their country is Nihon (or Nippon) Koku, meaning "land of the rising sun." It's an appropriate name because Japan is one of the first major countries to see the sun rise each day. Just as Japan leads the world into each new day, it has also been a world economic leader, although today it is a sputtering one.

Japan is an island country in the Pacific Ocean off the eastern coast of Asia. Geographically, the country is well positioned to access the markets both of Eastern Asia and across the sea in the Americas. Although Japan is the world's eighth-largest country in population with 126 million people, its 146,000 square miles make it only the fifty-ninth largest in area. The result is that Japan's population density (people per square mile) is among the highest on the earth. Japan is about the size of the state of California, but has almost four times as many inhabitants!

Song of the Samurai

Japan's earliest inhabitants were the Ainu, a basically Caucasian tribe who were conquered by successive waves of Asian invaders.

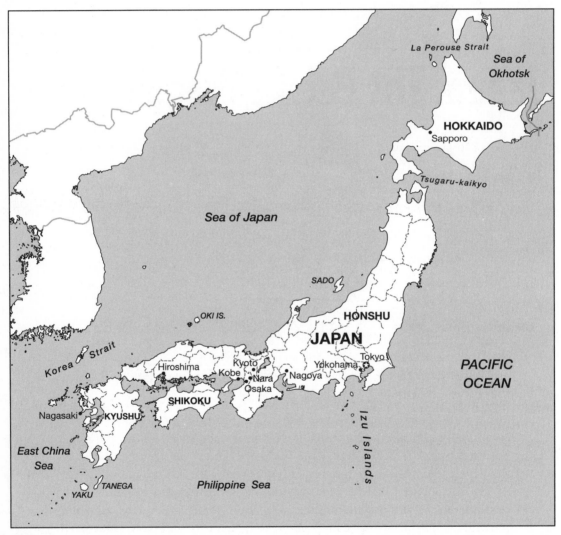

Japan.

The Chinese began to arrive on the Japanese islands about 400 C.E. They brought culture, writing, Buddhism, and, later, Confucianism.

As time went on Asians dominated the islands, and the Japanese people were united under a hereditary emperor. The emperors lost power to shoguns (warlord generals) during the Samurai period, a time of feudal strife, but came back under direct imperial rule under the Meiji Emperor (1868).

Although the Japanese islands had been insular and limited foreign contact, they quickly caught up by emulating the west. The Japanese soon became the prime movers in the Asian industrial revolution. Because the Japanese islands are poor in mineral deposits, especially coal, iron ore, and oil, the newly industrialized nation began to exploit some of its neighbors, particularly Korea and China, and entered into a period of military adventurism that ended in World War II.

The two atomic bombs dropped on Hiroshima and Nagasaki in 1945 made Japan the only nation in the world to be attacked with nuclear weapons, and this devastating new weapon brought a quick Japanese surrender.

Geographically Speaking

For more than 1,000 years the Japanese royal court lived in Nara and Kyoto and much of what is considered uniquely Japanese developed in these beautiful cities. Because neither city had war industries, the United States spared them from bombing in World War II. As a result, one of the largest bronze Buddhas in the world still placidly watches over Nara from the imperial deer park, and the more than 2,000 ancient shrines, temples, and gardens of Kyoto can still be enjoyed.

An Economic Giant Reborn from the Ashes

The United States occupied Japan from 1945 until 1952 and gave its defeated enemy massive financial and democratic aid. Japan quickly recovered and followed the United States in forming a strong, democratic, free-market economy.

A key notion seized on by Japan in its recovery was the cohesion of government, business, and labor. This would become the guiding principle of Japan, Inc.

Japan adapted a series of policies aimed at protecting domestic industries and encouraging export. Earnings were invested in research, and expansion was financed through low-interest, government-sponsored bank loans (made possible by high personal savings due to a deliberate scarcity of consumer goods). It took Japan until the mid-1950s to get back to prewar production levels. By the 1960s, its economy had skyrocketed. Growth rates throughout the decade averaged a phenomenal 10 percent a year. By the 1980s, growth had "slowed" to a more sustainable 5 percent a year, but Japan was now number one or two in the world in shipbuilding and in producing steel, automobiles, trucks and buses, watches, VCRs, TVs, cameras, microwave ovens, refrigerators, computers, and copying machines.

Terra-Trivia

Until recently, Japan was the second-largest consumer of electricity on the planet. With only 2 percent of the world's population, it consumes more than 7.5 percent of the world's total electricity. China has now jumped ahead of Japan. With 20 percent of the world's population, China consumes more than 9 percent of its electricity. The number-one consumer of electricity remains the United States, which, with 4.5 percent of the world's population, consumes more than 25 percent of the world's generated electricity.

In 1991, the Japanese economic juggernaut began to falter, as many countries to which Japan exported its goods were hard hit by recession. In addition, Japan's Asian neighbors and export partners South Korea and Taiwan had begun to produce higher-quality and lower-priced goods of their own. Japan felt increased pressure to open its own markets to lessen its imbalance of trade with other nations. All of these factors took a toll on Japan's economy. Throughout the 1990s, real-estate and stock-market values plummeted, thousands of small companies went bankrupt, and a series of financial and political scandals rocked the nation. A country that once seemed poised to rule the world economically now faces serious and troubling questions about its economic future.

Eye on the Environment

One of the defining factors of Japan's economy and recent history is its almost complete lack of natural resources. A whopping 61 percent of Japan's electricity is generated by imported fossil fuels (home-grown nuclear power accounts for 29 percent and hydro-power for another 9 percent).

But Japan is not sitting idly by. In a concerted effort to break its dependence on imported fossil fuels and meet its Kyoto Protocol commitments to reducing greenhouse gas emissions, Japan has become a leader in the solar revolution.

Land of the Rising Sun

Millions of years ago, tectonic action caused the Pacific plate to smash into the Asian plate, pushing up great mountains on the edge of the Asian continental shelf. A deep gorge formed where the Pacific plate and Philippine plate folded under the Asian plate. Today, the mountains are Japan, and the gorge is the Japanese Deep Trench (28,000 feet deep, off Japan's southeast coast). Japan's most beautiful mountain, Mount Fuji-san, is also its highest mountain (12,389 feet). This snow-capped volcanic cone was last active in 1707.

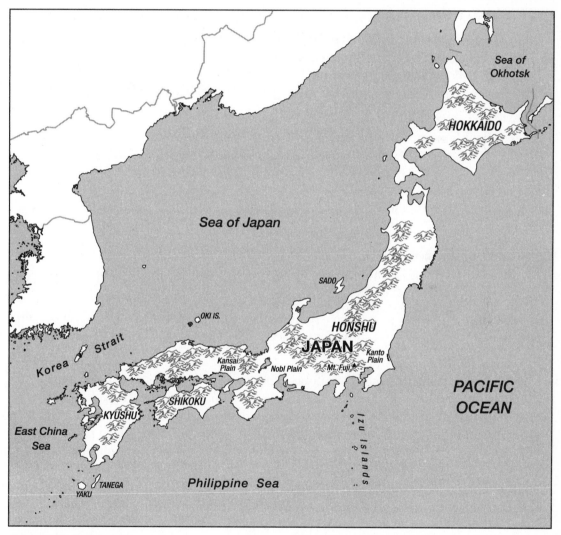

Japan—physical features.

The Waters and the Islands

Japan is an archipelago, which means that it's made up of many islands. Although the country has about 1,000 islands, you have to keep track of only the "big four," which loosely form the shape of a huge dragon. To the north is the dragon's head, the island of Hokkaido. Its northern latitude makes it a chilly place in the winter. Although Hokkaido is large, it's not as heavily populated as the islands to the south.

The next island south of the dragon's head (the body of the dragon) is Japan's largest island, Honshu. Considered to be Japan's mainland, it's home to Japan's largest cities and the bulk of its population and industry, especially in the Kanto Plain near Tokyo. Farther south, the other two large islands, Shikoku and Kyushu, form the tail of the dragon.

The Kuril Islands, which form a chain from Northern Japan to Russia, are an issue of great concern. Although Japan claims the islands as part of its national territory, Russia says "nyet"—it got them in a last-minute grab at the end of World War II and it occupies them. Even without the Kurils, Japan is almost 1,200 miles long and, at most, only 200 miles wide. No location in Japan is more than 100 miles from the sea.

Mountains and Plains

Japan has a major north-south spine of mountains called the Japanese Alps. The Japanese islands are 80 percent mountains and hills, and only 25 percent of Japan's land slopes less than 15 degrees. The islands are dotted with more than 200 volcanoes, 60 of which have been active in recent history.

Japan is on the Pacific's Ring of Fire (see Chapter 2) and is subject to volcanic eruptions and at least 15 earthquakes a year. The country also experiences a similar number of typhoons and at least one or two tsunamis (seismic sea waves) each year. Geologically, Japan is one of the world's most dangerous places to live. Tokyo sits largely on fill (land created artificially by filling in parts of Tokyo Bay) that could liquefy in the event of a major quake. Tokyo has narrow streets dotted with many skyscrapers built before stringent regulations were in force, and it's part of the world's largest population concentration. Japan's largest island, Honshu, has the world's largest urban conurbation—Tokyo, Kawasaki, and Yokohama. Japan's most dominant body of water is its Inland Sea (240 miles long and from 8 to 40 miles wide), connecting Honshu on the north with Kyushu and Shikoku on the south. The sea is known for its islets and beauty and is valued for its fishing.

Japan's Temperate Climate

Japan straddles the same latitudes as the east coast of the United States from mid-Maine to north Florida. The two areas have much in common. Because they're both temperate, they generally have similar temperatures and rainfall patterns, usually about 40 to 50 inches. Unlike the United States, though, Japan has two annual rainy seasons, with occasional coastal pockets getting as much as 100 inches of precipitation per year.

The People of Japan

Perhaps in no other developed country do the values and traditions of the past exert as much impact on the present as they do in Japan. The nation is extremely cohesive: it has one race (Mongoloid, similar to the Chinese and Korean people but smaller in stature), one language (Japanese), one culture, and one highly developed—and technology-oriented—system of education that produces virtually 100-percent literacy.

Until 1947, the state religion was Shinto, in which the emperor was worshipped as a god. Now freedom of religion is guaranteed to all. The Shinto religion, indigenous to the Japanese islands, involves the worship of the divine forces of nature. Ancestor worship also factors strongly into Shinto practice. Although most Japanese are still Shinto, some 16 percent are Buddhist. Most adhere to both religions because in practice the two largely overlap.

The U.S. occupation after World War II brought Westernization and transformation to Japan. Although patterns of rural life have changed little, it's in the cities, where 79 percent of Japan's people live, that change is most evident. The Japanese had a brief baby boom after World War II, but birth rates had decreased from 34 per thousand to 10 per thousand by 2002. With a drastic decrease in the death rate (life expectancy is now up to 79 years), Japan's population growth rate is only 0.3 percent, one of the lowest in the world.

The low birth rate is just one challenge facing Japan. Other challenges include its lack of resources; a banking system in serious need of reform; a growing scarcity of usable land for farming, industry, and people; and an aging population. Japan will soon face a labor deficit, which will be difficult to remedy given Japan's traditional aversion to immigration and reluctance to fully empower women in the workplace.

On the more positive side, the famous Japanese work ethic is still more or less intact, as is their leadership in the field of high technology. Japanese breakthroughs in robotics also hold great promise for the future. Japan is one of the world's principal economic engines, so it is in everyone's interest that the Japanese find a way out of their current dilemma.

The Least You Need to Know

♦ Japan is an island nation with one of the highest population densities in the world.

♦ Japan's use of its military might to acquire new resources and markets by force ended in catastrophic defeat at the end of World War II.

♦ In the aftermath of World War II, democratic reforms and an infusion of U.S. aid put Japan back on its feet and on the road to becoming an economic super-power.

♦ Japan today is the world's second most technologically powerful economy, but is struggling to snap out of more than 10 years of economic stagnation.

East Asia's Economic Tigers

In This Chapter

- ◆ A region unlike any other
- ◆ Colonial legacies
- ◆ Surveying the countries
- ◆ The Korean tinderbox

As its title indicates, this chapter is a little different in that it covers a region that is more economic than typically geographic. East Asia's four economic heavy-hitters, or Tigers, as they've been dubbed, are South Korea, Taiwan, Hong Kong, and Singapore. Although they do share certain geographic similarities—they are all located on the eastern fringe of Asia and they're all islands, peninsulas, or both—the reason for lumping them together is the strength and reach of their interconnected economies. Japan could have been included in this chapter as well, as it is closely tied with all four Tigers by trade and history.

Tracking the Tigers

As you can see on the following map, South Korea occupies the southern half of the Korean peninsula, just about 100 miles west of Japan. South

Korea is separated from China by North Korea, with which it shares the peninsula, and by the Yellow Sea. In fact, South Korea is actually twice as far from China as it is from Japan. The island of Taiwan (formerly called Formosa) lies off the southeast coast of China. Hong Kong occupies the tip of a peninsula and several islands in Southeastern China. At the southern tip of the Malay Peninsula, 1,500 miles southwest of China, is the island state of Singapore.

All four Tigers are highly industrialized and urban and enjoy high levels of economic development. As city-states, both Hong Kong and Singapore are virtually 100 percent urban. South Korea and Taiwan are about three fourths urban. Although the Tigers lost some of their roar in the Asian financial crisis of 1997–1999, the per capita gross domestic product (GDP) of Hong Kong and Singapore is the same as that of the most developed countries of Europe. The per capita GDP of South Korea is higher than that of European Union members Greece and Portugal; Taiwan's is just slightly below.

The four Tigers.

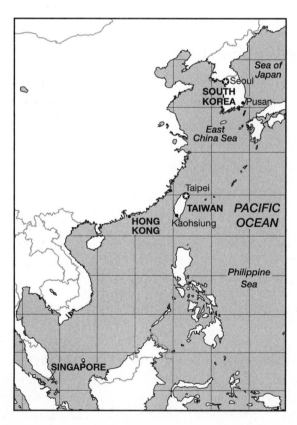

Individually, the four Tigers are the world's biggest exporters outside of North America, Japan, and the European Union. Each country annually exports between $120 billion (Singapore and Taiwan) to almost $200 billion (Hong Kong) worth of goods, mostly to the United States, China, Japan, and each other. In 2001, the four Tigers exported nearly $600 billion worth of goods, making them as a bloc the world's third largest exporter, behind only the United States and Germany but ahead of Japan, which came in fourth.

Because the islands and peninsulas of this "region" are not well endowed with natural resources, each country (or city, in the case of Hong Kong) must import raw materials, perform the necessary manufacturing and assembly operations, and then export the finished products. The four Tigers have some of the best, and best located, harbors in Asia and often serve as trade agents, shipping out other countries' goods. Throughout the "region," low-wage, labor-intensive industries are increasingly being outsourced and replaced by more technology-intensive industries.

Complex Intertwined Histories

All four tigers have been colonies at one time: Hong Kong and Singapore, of Britain; Taiwan and South Korea, of Japan. And all four were occupied by Japan during World War II. Both China and Japan have weighed heavily on the culture and development of South Korea, and their relationship with China is the dominant political and economic issue for both Taiwan and Hong Kong.

South Korea

Korea is an ancient land with a turbulent history. In 1910, Japan took over Korea as a colony and ruled it brutally until the end of World War II. After the war, the peninsula was split at the thirty-eighth parallel: the north went to the Soviet Union, and the south to the United States.

The northern half became the communist Democratic People's Republic (usually called North Korea), and the southern half became the capitalist Republic of Korea (usually called South Korea). In 1950, North Korea set off the Korean War by invading the south in an attempt to force unification. In 1953, a truce was signed and a no-man's land was set up just north of the thirty-eighth parallel, replacing the border between the two states. More than 50 years later, the participants in the Korean War have yet to sign a formal peace treaty. Since the war, South Korea has achieved extraordinary economic growth and prosperity, and fully integrated itself into the modern, high-tech global economy; it operates under basic democratic political ideals. North Korea continues to languish under a harsh Communist dictatorship and its economy is abysmally poor.

Geographically Speaking

Because there has never been a treaty ending the war, the Korean Conflict remains a hot spot of the old "cold war" between the forces of democracy and communism. As a result the United States has stationed 37,500 active-duty service men and women in South Korea since the end of open hostilities in 1953.

Taiwan

In 1949, communist forces under Mao Zedong defeated Chiang Kai-shek's Nationalist government in Mainland China, triggering an exodus of Nationalist forces to the island of Formosa, which was renamed Taiwan. Economic and military assistance from the United States enabled the Nationalist government to survive and begin to turn the island's economy around. For two decades, both the People's Republic of China (on the mainland) and the Republic of China (on Taiwan) claimed to be the true China. In 1971, however, with the assent of the United States, Taiwan was expelled from the United Nations and its seat given to the People's Republic of China. This has given the world the "Two China" solution, which recognizes Communist Mainland China, but also requires the tacit recognition of a free, democratic island nation of Taiwan (the other China).

Despite the political setback, Taiwan continued to make great strides economically. Although it is regarded by China as a renegade province and is officially recognized by few countries, Taiwan has trading relationships around the world. It is also a major investor throughout Southeast Asia and, despite ongoing tensions and conflicts, in China itself.

Hong Kong

In 1898, the British took over Hong Kong from the old imperial Chinese government, with a 99-year lease, and turned it into the best harbor/trading post between Shanghai and Indochina.

In the late 1940s, when the communist leader Mao Zedong was conquering all of China, he easily could have seized Hong Kong but chose not to because it was his gateway to the world for the goods his country desperately needed for economic survival. In 1984, as the lease was nearing expiration, the British agreed to give back all of Hong Kong in 1997 in return for a guarantee from China that it would preserve the economic system and civil liberties of its former colony.

In July 1997, Hong Kong became the Hong Kong Special Administrative Region (SAR) of China. Although it was hard-hit by the Asian financial crisis of the late 1990s and by the global slowdown of the early 2000s, Hong Kong remains one of the world's great trading economies. Although it is certainly in China's interest to maintain Hong Kong's economic autonomy and vibrancy, it remains to be seen how willing it will be to preserve the former colony's political freedoms.

Singapore

Singapore's modern history began with the establishment of a trading post on the Malacca island by the British in 1819. Singapore enjoyed two key advantages that helped turn it into one of the world's great ports: its strategic location at the mouth of the narrow passage connecting the Indian Ocean to the South China Sea and its tax-free status. With the opening of the Suez Canal in Egypt in 1869, Singapore became a major port of call for steamships taking the new shortcut from Europe to the Far East. Throughout the nineteenth and early twentieth centuries, immigrants from China, Malaysia, and India flocked to Singapore, drawn by its increasing prosperity.

In 1959, Singapore became a self-governing state, but four years later, in 1963, it joined with Malaya, Sabah (Northern Borneo), and Sarawak to form a new country, Malaysia. The union was short-lived. In 1965, Singapore separated from Malaysia and became an independent country and a member of the United Nations.

Terra-Trivia

Singapore is famous for its extreme (some would say oppressive) orderliness and tidiness. It is an immaculately clean, graffiti-free city, where people are encouraged to be kind, courteous, and quiet. Its crime rate is virtually zero.

The Big Players in the Game

South Korea and Singapore are separated by about 2,900 miles (approximately the width of the contiguous United States). The four countries and territories in this economic region are scattered over this vast expanse, each with its own politics, economics, and character.

South Korea

South Korea is blessed with the best agricultural land on the Korean peninsula; North Korea, on the other hand, struggles to feed its people adequately and has experienced catastrophic food shortages in recent years. The bulk of the peninsula's raw materials, however, are in the north. It is one of the many tragic ironies of the continuing division of the Korean peninsula that the two halves with their complementary resources cannot find common ground to work together.

Although South Korea is known as a major producer of semiconductors and electronic products, its industrial output also includes passenger cars (Hyundai), cement, pig iron, crude steel, chemicals, and textiles. South Korea is also one of the world's largest shipbuilders. Like everyone else in this region, South Korea was shaken by the Asian financial crisis of the late 1990s, but it has weathered the global economic slowdown of 2001–2002 well, thanks to the continuing strength of its industrial and construction sectors.

South Korea's capital is Seoul. Between one fourth and one third of the entire population of the country lives in or around this capital city. The next-largest and next-most-important city is Pusan, South Korea's largest port city and also a major manufacturing center.

Taiwan

Taiwan's export-driven economy has generated a significant trade surplus and one of the world's largest supplies of foreign reserves. Although Taiwan's modern economic development was based on the production of steel, machinery, paper, cement, and textiles, today the emphasis is increasingly on high-tech items, such as personal computers, telecommunications equipment, radios, TVs, VCRs, and calculators. Most of these goods go to the United States, Hong Kong, Japan, and Europe. Hong Kong (and indirectly China) accounted for 21 percent of Taiwan's exports in 2001. Despite continuing tensions over Taiwan's status in relation to China, in 2001, Taiwan eased long-standing restrictions on direct investment in China. It is estimated that there are some 50,000 Taiwanese businesses operating in mainland China.

Taiwan's capital, Taipei, is also its preeminent city. More than 6.5 million people live in the Taipei metropolitan area, on the northern part of the island. The second-largest city is the southwestern port of Kaohsiung.

Hong Kong

Hong Kong became part of China once again in 1997. Under the "One Country, Two Systems" formula agreed to by China and the United Kingdom before the handover of Hong Kong, the territory retains its economic and political autonomy in all things except foreign affairs and defense policy. Whether China can refrain from imposing its political and judicial system on Hong Kong remains to be seen, but at least for now, Hong Kong is, for all intents and purposes, a separate entity from China.

Hong Kong recovered quickly from the Asian financial crisis of the late 1990s, but has been severely affected by the more recent global slowdown. Still, it retains one of the world's highest per capita GDPs.

Singapore

Space, or a lack of it, is a key to understanding Singapore. Much of its food and most of its water must be imported. Most housing is high-rise, and Singapore doesn't have enough of it, especially for middle- and low-income families. Yet, Singapore has a strategic location and superb harbor, so it has developed as a major *entrepôt*.

Crude oil from Southeast Asia is imported, refined, and then transshipped to other Asian countries. Malaysian products (rice, spices, and timber) are also exported through Singapore, as are Malaysian and Indonesian rubber. The list of products, goods, and raw materials shipped through Singapore by less-developed neighbors with poorer port facilities is endless. Singapore also imports products for transshipment to other Asian countries (machinery and autos, for example).

def•i•ni•tion

An **entrepôt** is a port that specializes in the transshipment of goods from one country (not its own) to another country (or other countries). Hong Kong plays this role, and Singapore does it in spades.

Electronics and financial services top the list of Singapore's domestic industries. Others include shipbuilding and repair, oil refining, rubber processing, food processing, chemicals, pharmaceuticals, plastics, and clothing. Its main export partner is Malaysia, followed by the United States, Hong Kong, Japan, Taiwan, Thailand, China, and South Korea. Singapore has also become an important financial center with a prominent stock market.

The Lay of the Land

Although South Korea and Taiwan are full-size countries with varied terrain and large agricultural areas, Hong Kong and Singapore are essentially cities with very little rural land.

South Korea

In area, South Korea is by far the largest country in this region—it's almost three times the size of the next-largest country, Taiwan. Even so, it's only about the size of Indiana. South Korea is a rugged, mountainous country, especially on its eastern side. As you might guess, because of the mountains, most of the population of more than 48 million tends to be distributed on the western side of the peninsula.

South Korea has cold winters and warm-to-hot summers; the waters surrounding the peninsula have a moderating influence on South Korea's climate. The country gets most of its rain on its southern coast, where agricultural land is most plentiful. Because of the temperate climate, more than one crop per year can be grown, and rice cultivation is of primary importance. South Korea is also vulnerable to an occasional typhoon (hurricane-like) deluge.

Taiwan

The island of Taiwan is only a little more than 100 miles off the southeast coast of China. Located on the edge of the Philippine tectonic plate, it's extremely vulnerable to earthquakes. And as if that weren't enough, Taiwan also gets its share of tropical typhoons.

Although Taiwan measures only 13,900 square miles, it has almost 22 million people. Hilly and mountainous in the east, it has a long coastal plain in the west. The Tropic of Cancer crosses the island, which means that it's at the edge of the tropics. The result is minimal seasonal temperature variation—Taiwan has hot summers and warm winters. Rainfall is heavy throughout the year.

Because of the country's warm temperatures, multiple cropping is practiced on the western half of the island. The fields are intensively farmed, largely in small subsistence plots. The dominant crop in Taiwan, as in South Korea, is rice. Tropical rain forests grow on the eastern, mountainous side of the island.

Hong Kong

Hong Kong is located at approximately the same latitude as Southern Taiwan, on China's southeastern coast. The former British territory is made up of the New Territories and Kowloon on the tip of a peninsula and several islands, chiefly the island of Hong Kong, scattered across one of Asia's busiest and best harbors.

Although Hong Kong is in the tropics, the southwestern monsoon makes the weather subtropical. This monsoon is a warm, moist equatorial wind that gives Hong Kong a rainy season in early summer. The temperature range is 59–82 degrees F. The area is naturally hilly and rugged.

Singapore

Singapore is a city-state, a republic, and an island off the southern tip of the Malay Peninsula. Although Singapore is one of the largest ports in the world, it's the smallest of the Pacific Rim states in both size (only about 250 square miles) and population (3 million). If Singapore were in Europe, it would vie for microstate status.

Unlike the other countries in the Pacific Rim, Singapore is not mountainous. Despite years of dredging, in fact, it's still somewhat swampy. Because Singapore is only one degree north of the equator, it lies in the heart of the tropics; because it's almost on the equator, it's tropical, wet (averaging 95 inches of precipitation annually), hot (81–86 degrees F), and humid.

The Ethnic and Religious Mix

South Korea's people are 99.9 percent Korean, and all speak Korean. More than half are atheists, one quarter are Buddhists, and about one fifth are Christians. More than three fourths of South Korea's people are urbanites.

In Taiwan, most of the people live in the north and the west. Highly urban, the bulk of the population is Taiwanese (84 percent), and almost all the rest are mainland Chinese, either Nationalists who fled from China in 1949 or their descendants. Mandarin Chinese is the official language. The main religion is a mix of Confucianism, Taoism, and Buddhism; about 5 percent of the population is Christian.

The people of Hong Kong are 97 percent Chinese, and English and Cantonese Chinese are its official languages. The prevalent religion is a blend of Confucianism, Taoism, and Buddhism; about 8 percent of Hong Kong people are Christian.

The smallest of the group in area, Singapore is the most ethnically diverse. About 78 percent of the population is Chinese; 14 percent, Malay; and 7 percent, Indian. Interestingly, Singapore has four official languages: English (the language of administration), Mandarin Chinese, Malay, and Tamil. Its primary religions are Buddhism, Islam, Hinduism, Confucianism, Christianity, and mixtures of all five.

Terra-Trivia

If an armed conflict with North Korea were to break out, South Korea's capital and its huge population cluster would be extremely vulnerable. Seoul is only about 25 miles from the demilitarized zone (or DMZ), which marks the border between North Korea and South Korea. Although this zone is one of the most heavily defended lines on Earth, the threat is grave and real. Since the North Korean government detonated an atomic bomb in October 2006, the situation has become even more volatile.

The Least You Need to Know

- South Korea, Taiwan, Hong Kong, and Singapore are four small, very industrialized trading states located on the eastern fringe of Asia. Because of their economic strength and prowess, they have been called the East Asian Tigers.

- Taken together, the four Tigers are the world's third-largest exporter.

- The Republic of China on Taiwan owes its birth to the victory of the Communist Revolution in China, which forced the Nationalists to flee to the island then known as Formosa.

- Hong Kong's future depends on how well its union with China goes and whether China keeps its promise to maintain a "One Country, Two Systems" policy.

- Singapore is the smallest country in the region, a very prosperous city-state whose economy is based on export and transshipping.

Chapter 14

Australia and New Zealand: Looking Down Under

In This Chapter

- ◆ Exploring the land of volcanoes
- ◆ Trekking in the outback
- ◆ Aborigines and Maoris were here first

Australia and New Zealand, which are neighbors despite being separated by 1,300 miles of Tasman Sea, form a region in the far South Pacific. Both countries are firmly based in English tradition, language, and rule of law and government, despite their 12,000-mile distance from Europe. Young relative to many other nations, literate, and prosperous, New Zealand and Australia share many similarities.

The people of this area are serious, hard-working, independent souls, not to be pushed around. The region is also serious about its "no-nuke" policies. The 1985 Treaty of Rarotonga declared the South Pacific to be a nuclear-free zone.

This region contains a number of islands, but only two countries, one of which encompasses an entire continent. Although Australia is by far the larger of the two in both area and population, let's start with the little guy, New Zealand.

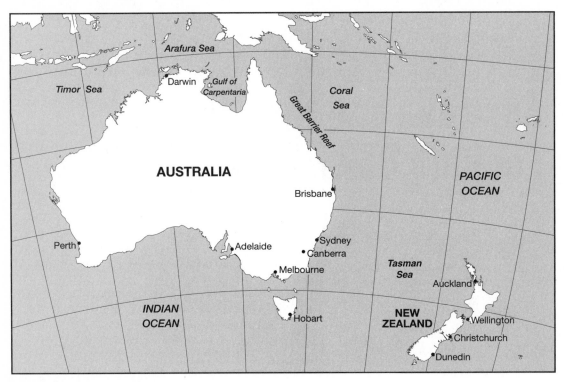

Australia and New Zealand.

Land of the Kiwi: New Zealand

If New Zealand were placed in the Northern Hemisphere, it would be at approximately the same latitude as the stretch from Los Angeles, California, to Seattle, Washington. The country is farther south than Australia and cooler—closer to the South Pole. New Zealand's Stewart Island hosts lots of penguins.

First Peoples: The Maoris

New Zealand's first settlers were the ancestors of today's Maoris, a Polynesian people who began arriving in the islands as early as the ninth century C.E.

The Flood of Europeans

Dutch, English, French, and even American explorers and whalers visited New Zealand, but it was eventually the British who seized large areas of the islands and began to extensively settle, even though the Maoris violently resisted. There are still land disputes between the "white" settler population and the indigenous Maori population of these beautiful islands. Today, New Zealand is an independent nation, governed by an elected prime minister, but recognizes the British monarch as head of state.

Trembling Land

New Zealand is more than 1,000 miles long and about 1,250 miles southeast of Australia. The country is sandwiched between the Tasman Sea on the west and the South Pacific on the east. It's a mountainous nation consisting of two main islands, North Island and South Island. (A third island, Stewart, located south of South Island, is significantly smaller than the two main islands.)

Although dwarfed by Australia in size, New Zealand is still slightly larger than the United Kingdom, the European source of most of its population. Separating the two main islands is Cook Strait, which thins to 16 miles at its narrowest point. North Island is slightly smaller than South Island but is more geologically active.

Unsteady Ground on North Island

Volcanically speaking, North Island is active indeed. There are three major volcanoes in the island's center, and one of them, Ngauruhoe, erupted in 1995. All three volcanoes are part of Tongariro National Park. To the west in Egmont National Park, extinct Mount Taranaki (Egmont) rises more than 8,000 feet above the rich surrounding plains and vineyards.

With geysers shooting 100 feet high and fumaroles (a surface hole from which volcanic gases gush out) spewing steam and gas, New Zealand derives about 7 percent of its power from geothermal plants. South Island's rushing streams

Eye on the Environment

Like other Pacific Ring of Fire nations, New Zealand has abundant geothermal resources, most of which remain untapped. It has been estimated that using current technologies about 75 percent of the country's peak electricity demand could be met by high-temperature geothermal sources. Today only a fraction of that potential has been developed.

are also used to generate hydroelectric energy, which is transmitted by undersea cable across Cook Strait to the more heavily populated North Island.

In the heart of North Island is Lake Taupo, the island's largest lake, teeming with fish. To the far north is a long peninsular proruption (long extension of land) called the Northland. This mostly low-lying area is famous for its 90-mile beach north of Auckland.

New Zealand—physical features.

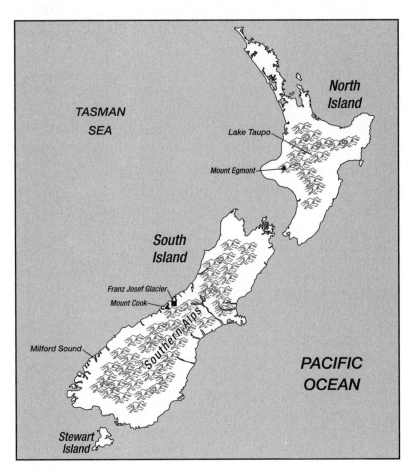

The Alpine South

With a surface area of 58,000 square miles, South Island is larger than North Island and has a more varied terrain. Its dominant physical feature is the Southern Alps (Fjordland), a massive mountain chain that runs along the west coast of the island, providing spectacular alpine scenery and 16 miles of uninterrupted ski runs famous the world over. New Zealand's highest mountain, Mount Cook, or Aorangi, rises 12,316 feet near the center of the Southern Alps.

The Southern Alps are cloaked in glaciers that shrink and expand depending on weather patterns and the rate of snowmelt. The Franz Josef Glacier is unique in that it ends in a temperate rain forest. In the southwest portion of the island are Milford Sound's Mitre Peak and its matching fjord. Here, 6,000-foot mountain faces rise from the water, high valleys unleash thundering falls (Sutherland Falls is fourth highest in the world at more than 1,900 feet high), and seals play on the shore.

Geographically Speaking

The kiwi is a small, chicken-sized, flightless bird found only in New Zealand. Kiwi birds have long, slender beaks, which are excellent for finding and digging out their diet of worms and grubs. The nocturnal kiwi is endangered, and all five varieties have severely reduced numbers in their home ranges. During World War I, some New Zealand soldiers, fighting for Great Britain, carved a giant kiwi image in the chalk hill outside Sling, England. From then on, the popular name for New Zealanders has been "kiwi," after their national bird.

South Island's west coast was the site of the 1860s kiwi gold rush, which at its peak in 1866 produced more than 700,000 ounces of bullion. The area is also a rich source of greenstone, or jade.

The Kiwis

North Island is home to three quarters of New Zealand's more than 4 million people, including most of the country's indigenous Maoris. North Island also boasts the country's largest city (Auckland), its capital (Wellington), three quarters of its industry, most of its dairy production, and more than half of its huge sheep population.

Because of the mountainous nature of New Zealand's interior, the majority of its population lives in coastal cities and towns (where most of the country's rail lines and roadways are also found). Nearly 80 percent of the population is of European descent, while Maoris account for about 15 percent. English is the most widely spoken language, but both English and Maori are official languages.

Protestant religions prevail in New Zealand, although a sizable minority of New Zealanders (about 15 percent) are Roman Catholic. New Zealand's Catholic population is largely composed of descendants of Irish immigrants who arrived in the country in the 1800s and brought their religion with them.

The Maori

New Zealand's indigenous Maoris are unrelated to Australia's indigenous people. Maoris are a Polynesian group while Australia's Aborigines are an Australoid group. With the arrival of the pakeha (white man), the Maori population shrank from an estimated (and disputed) high of 250,000 to about 40,000 in 1900, thanks in large part to diseases and firearms introduced by the European settlers. Maoris have enjoyed parliamentary representation since 1867 and their population recovered during the course of the twentieth century. Today, the Maoris live mostly in cities and have integrated fairly successfully in New Zealand society while still retaining their cultural traditions and identity.

Livelihoods and Pastimes

Agriculture and food exports are major components of New Zealand's economy. The fertile Canterbury Plains on the eastern side of South Island are the country's breadbasket. It has enough good land to grow 90 percent of the country's grains. Hops, apples, and raspberries thrive.

The economic restructuring that began in the mid-1980s has helped transform New Zealand into a more industrialized and service-oriented free-market economy. Major industries include food processing, the manufacture of wood and paper products, textiles, machinery, banking, and tourism.

Terra-Trivia

Nearly 30 percent of New Zealand is forested, while more than 50 percent is given over to sheep and cattle grazing and farming. (New Zealand has about 40 million sheep, roughly 10 times its human population.) About 30 percent of the nation's total area is publicly held protected land administered by the Department of Conservation.

New Zealand's Cities

New Zealand is highly urbanized: 85 percent of its people live in the cities. The country's largest city by far, Auckland proper has about 380,000 people, while the greater

Auckland area numbers more than a million people. The population of this bustling North Island city includes the largest percentage of Polynesians and Asians in New Zealand.

Wellington is New Zealand's capital, seat of government, and second-largest city. Located on the southwestern tip of North Island, Wellington straddles a fault, much like San Francisco, and is similarly earthquake-prone. The largest city on the South Island is Christchurch, known as New Zealand's "English City" as well as its "Garden City."

The Land Down Under: Australia

Australia's earliest inhabitants, the distant ancestors of today's Aborigines, reached the continent between 40,000 and 70,000 years ago by traveling over the land bridges and shallow seas that connected Ice Age Asia to Australia. Like other Stone Age peoples, Aborigines lived and hunted in isolation, using stone (not metal) tools.

Like New Zealand, the English settled Australia, but the earliest did not come voluntarily. Most of Australia's pioneering settlers were convicts sent over by the British government to relieve overcrowding in its prisons. Free settlers soon followed the convicts and took up farming, mining, and trading. The introduction in 1797 of a hardy Spanish breed of sheep with superior wool had a huge and positive impact on Australia's economic future. Farmers pushed into the interior across Australia's eastern mountains to "squat" on land in view of eventually taking possession of it. Displaced Aborigines were forced even farther into the interior as bloody conflicts escalated over land rights and access to scarce food supplies.

Gold was discovered in New South Wales in the 1850s, and the resulting gold rush helped swell Australia's sparse population. Friction with Chinese workers in the gold fields in 1861 led to immigration restrictions and a "white Australia" policy. After World War II, Australia benefited significantly, both economically and culturally, from an influx of non-British European immigrants, while in the 1980s it took in large numbers of Southeast Asian refugees, especially from Vietnam and Cambodia. Australia today has become a multicultural melting pot: half of all Australians are foreign-born or have a foreign-born parent.

The Big Dry: Australia's Land

Although Australia is a huge place (the world's sixth-largest country), it's the world's smallest continent. Roughly the size of the United States minus Hawaii and Alaska,

Australia is the land of superlatives. It's the lowest and the flattest continent. Except for Antarctica, Australia is also the driest continent and has the sparsest population. Covering nearly 3 million square miles, it's surrounded by one sea after another (Coral, Tasman, Timor, and Arafura) and the Pacific and Indian oceans, as shown on the following map.

Mountains and Outback

Australia's main physical regions are the Great Western Plateau, the Central Lowlands, and the Eastern Highlands. The Western Plateau, with no trees or rivers, but lots and lots of sand, encompasses about two thirds of Australia. Sand ridges that reach 60 feet high, sand hills, sand dunes, and sand plains make up the Great Sandy Desert, the Gibson Desert, and the Great Victoria Desert, which cover the bulk of central and Western Australia. Central Australia is also the site of the country's most famous natural landmark, Uluru, or Ayers Rock.

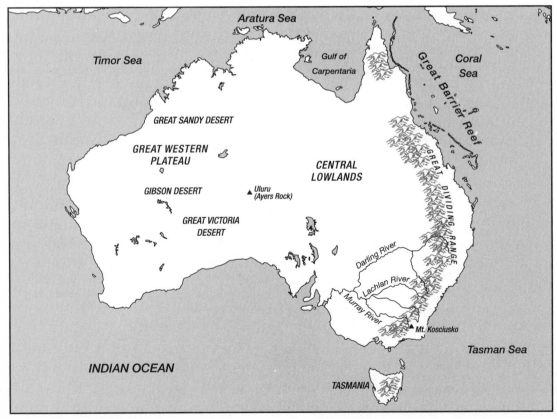

Australia—physical features.

The Central Lowlands extend from the Gulf of Carpentaria (the big notch in the north) through the pastureland of the Great Artisan Basin. The area is laced with intermittent streams that sometimes swell with the help of runoff from mountains to the east. Much of this water is too salty for crops and is used only to irrigate grazing land. The best farmland is located in the southeast, where the Murray, Darling, and Lachlan rivers flow; sediment in this area has also yielded opals.

The Eastern Highlands, or Great Dividing Range, separate the dry interior from the fertile coast. The area is tropically sticky and hot in the north and humid and subtropical in the south. Steep mountains (such as 7,310-foot Mount Kosciusko, Australia's highest) rise in the southeast and are the source of a great deal of hydroelectric production. The mountains in the south feed the long, gentle rivers that water the area and make possible the "fertile crescent" filled with apple and pear orchards, farms, and sheep stations.

Tasmania, Australia's island to the south, lies 150 miles across the Bass Strait and is the mountainous southern extension of the Eastern Highlands. Dairy farming, sheep grazing, oil drilling, and hydroelectric power generation are big industries in Tasmania. Native wildlife species on this wet and isolated island include wombats, of which few remain; the Tasmanian wolf; and most famous (thanks to its cartoon alter ego), the threatened Tasmanian Devil.

The Great Barrier Reef

The Coral Sea off Australia's northeast coast contains one of the world's most splendid natural wonders: the Great Barrier Reef. It consists of 600 islands and atolls that extend some 1,250 miles in shallow water 10 to 100 miles off shore. The size of England and Scotland combined, this immense living mass of coral reef, the earth's largest, covers 80,000 square miles. It's the home of 400 types of coral and 1,500 varieties of fish. Three million terns and boobies cluck and squawk above, and green sea turtles, sharks, and giant clams live below.

The Great Barrier Reef Marine Park was established in 1981 in an attempt to save the marine life of the reef from the depredations of (turtle soup) canning factories, oil drilling, bombing practice by the Australian navy, and reckless tourists. The park is the world's largest protected marine area.

Australia's Climate, Flora, and Fauna

In Australia, south is cold and north is hot. The top third of the country is tropical, and the seasons are opposite of what they are in the Northern Hemisphere. Tasmania and the south coast of the mainland have a marine climate. Elsewhere, Australia's climate is fairly mild—perfect for tennis and surfing all year. Most of this vast country is arid or at least semiarid.

The "land down under" supports an amazing variety of plants that grow only in Australia—honey flower, spear lily (12 feet high with red blossoms), 600 kinds of orchids, aromatic eucalyptus trees, more than 500 types of gum trees, and acacia, the sweet-smelling plant known as the "wattle tree." (Settlers used acacia saplings for wattling, or interlacing, the frame for mud and plaster walls.)

Australia's wildlife is also quite unique. Kangaroos and other marsupials make up half the continent's native mammals. Marsupial females carry their young in a pouch until the babies are fully developed. Many types of kangaroos hop with surprising speed across the open plains. An endearing symbol of Australia, the cuddly koala bear is a marsupial and not a bear at all. It spends its time clinging to eucalyptus trees, its only food source. But the prize for oddest of all Australian animals goes to the duck-billed platypus, a web-footed mammal that lays eggs. (New Zealand, on the other hand, can claim only the bat and eel as native critters since the huge 10 to 13 feet high, 500-pound flightless bird known as the Moa was hunted to extinction centuries ago by the ancestors of the Maori.)

People of the Southern Continent

Australians are extremely urbanized: 85 percent of them live in the cities, which are located primarily along the coasts. The same population distribution pattern occurs in New Zealand, but there it was the mountains that discouraged interior settlement; in Australia, it was the inhospitable desert.

Australia's population is almost completely literate and English-speaking. About one quarter of the population is Anglican, one quarter is Roman Catholic, and another quarter is of mixed Christian faiths. Just as with New Zealand, Australia's population is largely distributed along the coasts. The difference is that in Australia, it is not mountains but the arid nature of the interior that explains the coastal distribution. Aussies have a prosperous economy and enjoy a high standard of living, with a gross domestic product per capita of over $32,000, slightly higher than the four dominant European economies (Germany, the United Kingdom, France, and Italy).

The arrival of British settlers in Australia proved a disaster for the Aborigines. Their numbers dwindled due to the introduction of European diseases and conflict with the settlers. Those who survived were relegated to menial subsistence jobs and doomed to a life of poverty and ill health. Conditions are still difficult for Aborigines in Australia, although there have been improvements and attempts to redress past wrongs. Full citizenship was finally granted to Aborigines in 1967, segregated schools are closed, and welfare benefits have increased. Many Aborigines live and work as graziers (sheep farmers) near sheep or cattle stations in the willie woolies (outback, or back country) or in the mines. Those living in towns tend to speak their indigenous language (or a creole mix) and cling to tribal ways. Alcoholism and drug abuse are serious and widespread problems in Aboriginal communities.

The Australian States, Territories, and Cities

Australia is divided politically into six states and two territories. To the west is—you guessed it—Western Australia, in north central Australia is Northern Territory, and south of that is South Australia. In the northeast is Queensland, and south of that, in the southeast, is New South Wales. To the far south is Victoria. Off the southern coast is Tasmania, and the small governmental area carved out of New South Wales is called the Australian Capital Territory.

Australia is a country of city dwellers. Its capital, Canberra (population: 330,000), is the only major Australian city not located on the coast. With a population of 4.2 million, Sydney is Australia's largest city and a major tourist destination with more than 70 beaches. Its spectacular harbor is dominated by the world-famous Sydney Opera House. Melbourne, located at Australia's southern tip across the Bass Strait from Tasmania, is the country's second largest city (population: 3.6 million), followed by Queensland's lush, subtropical capital, Brisbane, home to some 1.8 million people. Australia's Indian Ocean gateway is the thriving, if still somewhat remote, port city of Perth with some 1.4 million residents.

The Least You Need to Know

 ◆ Both Australia and New Zealand have indigenous populations, the Aborigines and Maoris, respectively.

 ◆ Both countries share an English heritage of language, law, and government.

 ◆ Australia and New Zealand are isolated by distance from the rest of the world but are not isolationist.

◆ The population of both countries is highly urbanized and distributed mainly along the coasts.

◆ Australia has extremely varied flora and fauna that, because of the country's remote location, are found nowhere else in the world.

Part 3

A Regional Look: Nations in Transition

All of the countries you'll visit in Part 3 share one very important characteristic: They either had or still have a communist political system. And with varying degrees of success, most of them are making the difficult but critical transition from a government-controlled socialist economy to a more open, free-market capitalist economy.

Each of the three regions covered in this part—Eastern Europe, Russia, and China—reflects a different stage in the transitioning process. Many of these countries—especially those in the European Union—have moved into the developed-world category. Others, however, have a tougher challenge ahead.

"Truth is, I don't know anymore myself. Ends in -stan or -okia."

Chapter **15**

The New Eastern Europe

In This Chapter

- ◆ The Soviet legacy and new lines on the map
- ◆ Joining the EU: the new Eastern European member states
- ◆ No longer in the USSR: Belarus, Ukraine, Moldova
- ◆ European flashpoints: the former Yugoslavia and Albania
- ◆ Understanding the physical features of Eastern Europe

Eastern Europe is a collection of "formers": former Eastern bloc countries that were, until recently, satellites of the former Soviet Union; former Soviet republics that, until recently, were portions of the huge country of the Soviet Union; former pieces of Yugoslavia that, until recently, were united under a single totalitarian government. Eastern Europe has undergone drastic changes since the collapse of the Soviet Union in 1991. More changes recently occurred in 2004 when some of them joined the European Union, with two more joining in January of 2007.

A History of Turmoil

This area is a battleground between the larger, stronger nations of Western Europe (think Germany in the two world wars) and the territorial ambitions of Russia. They were the central battlegrounds during much of the Nazi turmoil.

Behind the Iron Curtain

Following World War II, the victorious Soviet Union imposed communist regimes on much of Eastern Europe. Economically, ideologically, and militarily, the Soviet Union dominated its so-called *satellites:* East Germany, Poland, Czechoslovakia, Hungary, Romania, and Bulgaria. Under its strong-willed leader Marshall Tito, Yugoslavia managed to resist Soviet domination and go its own way. The small and isolated country of Albania also embraced communism, but alone among the communist states of Eastern Europe, it aligned itself with Maoist Red China instead of the Soviet Union.

def•i•ni•tion

A **satellite**, in the political sense, is a country subordinate to, or dominated politically or economically by, another country.

The defeat of Nazi Germany in World War II and the redistribution of the territories it had conquered resulted in a dramatic redrawing of boundary lines throughout east and central Europe.

As the cold war intensified in the 1950s and 1960s, several Eastern European countries attempted to gain greater autonomy from the Soviet Union. Uprisings and general strikes in Hungary (1956), Czechoslovakia (1968), and Poland (1960s to 1970s) were quelled by the Soviets, sometimes through brute force. Eastern Europe foundered economically under the communist system.

Nations Reborn

In December 1991, the old Soviet Union broke apart into 15 independent republics, 6 of them in Eastern Europe (7 if you count Russia itself). In the northern part of Eastern Europe, the Baltic states of Estonia, Latvia, and Lithuania regained their independence. In what had been the western edge of the Soviet Union, bordering the former satellite countries, the newly independent states of Belarus (or Byelorussia or Byelarus), Ukraine, and Moldova emerged. Among the satellites, East Germany rejoined West Germany and in 1993 the former Czechoslovakia split peacefully to form the Czech Republic and Slovakia.

Eastern Europe.

Yugoslavia had always been rife with territorial competition and ethnic tension. By the early 1990s, Yugoslavia was splitting apart along ethnic lines. Although Slovenia seceded relatively easily in 1991, the rebirths of Croatia, Bosnia and Herzegovina, Macedonia, Serbia, and Montenegro were painful and violent.

Most of the nations of Eastern Europe have made remarkable strides since the late 1980s in adapting Western-style democracy and free-market capitalism.

Joining the West: The New EU States

The new Eastern European member states of the European Union include three former Soviet Republics—the *Baltic states* of Estonia, Latvia, and Lithuania—and five former Soviet satellite countries—Poland, the Czech Republic, Slovakia, Hungary, and Slovenia. Two other former satellite states, Romania and Bulgaria, joined the EU in January of 2007.

Estonia

The smallest of the Baltic states, Estonia enjoys the highest per capita gross domestic product of the three. It is a flat and scenic country dotted with lakes. Its capital, Tallinn, is home to more than half a million residents, or nearly one third of Estonia's population. Because of the large number of Russians living in Estonia (about 28 percent), Russia still has a stake in the country and has yet to sign a 1996 border agreement with its small neighbor.

def•i•ni•tion

The **Baltic states** take their name from their position on the Baltic Sea, a long arm of the Atlantic Ocean by way of the North Sea. Poland, Germany, Denmark, Sweden, and Finland also share a coastline on the Baltic Sea.

Estonia's rich oil-shale reserves make it self-sufficient in energy, but exploiting those reserves took its toll on the environment, especially during the years of Soviet rule. Estonia's environmental situation has been improving steadily, however, with air-pollution emissions down considerably since 1980.

Only the narrow Gulf of Finland separates Estonia from Finland, and not surprisingly, the two nations have a great deal in common. The Estonian language is related to Finnish; along with Hungarian, they are Uralic languages, not Indo-European. As in Finland, Lutheranism is the dominant religion in Estonia, followed by Orthodoxy.

Latvia

Like Estonia, Latvia has a large Russian minority (about 30 percent of the population) and unresolved border issues with Russia. It also has substantial Ukrainian and Belarusian minorities. With more than 790,000 residents, the Latvian capital Riga, sometimes called "the Paris of the Baltic" for its striking urban landscape, is the largest city in the Baltic states.

Latvia is heavily dependent on imported energy and raw materials, but about 70 percent of its electricity is generated by domestic hydroelectric and thermal plants. The growing importance of the service sector in Latvia's economy has had the added benefit of easing many of its Soviet-era environmental problems.

The country's official language is Latvian, a Baltic language related to Lithuanian. English, German, and Russian are widely spoken as well. Lutheranism, Catholicism, and Russian Orthodoxy are the main religions.

Lithuania

Lithuania is the southernmost Baltic state and, with a population of about 3.7 million, the most populous. Its capital, Vilnius, has some 600,000 residents. Through the centuries Lithuania and its neighbor to the south, Poland, forged close cultural ties with each other. Like Poland, Lithuania is heavily Roman Catholic.

This gently rolling, forested land has a strong agricultural base. Lithuania's trade is increasingly oriented to the West rather than to Russia, and it has largely completed the process of privatizing its enterprises in preparation for joining the European Union. Lithuania is famous for its amber, the fossil resin that resembles a semiprecious stone.

Lithuania is more ethnically homogeneous than either Latvia or Estonia. The Lithuanian language is a Baltic language closely related to Latvian. Polish and Russian are also widely spoken.

Poland

Poland is the largest of the six former Soviet satellites in land area (more than 120,000 square miles) and population (38.6 million). Much of Poland is a vast plain with the only highlands in the south where the Carpathian Mountains run through the country's southern border with Slovakia.

Poland is an ancient country with a proud if tormented history that has seen it swallowed up by neighboring powers on several occasions. The rise of the Polish independent trade union Solidarity in the 1980s played a key role in the fall of communism in Eastern Europe and Russia. Reforms in the early 1990s revitalized Poland's economy and helped turn it, for a time, into one of the strongest in the region. Poland joined the European Union in 2004. Heavy industry remains a major, if weakening pillar of the economy, and a hefty 27 percent of the workforce is involved in agriculture.

The Poles speak Polish, a Slavic tongue, and are overwhelmingly Roman Catholic in their faith. Warsaw, with a population of 1.6 million people, is Poland's capital, largest city, and cultural heart.

The Czech Republic

At the end of World War I, the closely related Czech and Slovak peoples joined to form the country of Czechoslovakia. After World War II, a territorially reduced Czecho-slovakia became a Soviet satellite. In 1968, a Warsaw Pact invasion put an end to the famous Prague Spring, Czechoslovakia's brief but brave attempt to liberalize com-munism. The country finally attained its freedom in 1989, and in 1993 peacefully split along ethnic lines to form the two new countries of the Czech Republic and Slovakia.

> ### Geographically Speaking
>
> The cold war pitted two enormous military alliances against each other. The Warsaw Pact was a mutual defense alliance created by the Soviet Union in 1955 in response to the perceived threat from the North Atlantic Treaty Organization (NATO), the mutual defense alliance of the Western democracies created in 1949. The Warsaw Pact nations included the Soviet Union and all of the Eastern European communist countries except for Yugoslavia. After the collapse of the Soviet Union, NATO has expanded to include several former Warsaw Pact nations.

The Czech Republic is the more urbanized, industrialized, and Westernized part of the former Czechoslovakia. It is also one of the most economically stable and prosper-ous countries of Eastern Europe, second only to Slovenia in per capita gross domestic product. The heart of the republic is the splendid capital of Prague (population 1.2 million), one of Europe's big tourist magnets. The people speak Czech, a Slavic lan-guage close to Slovak and Polish. Roman Catholicism is the predominant religion in the republic, although acknowledged atheists slightly outnumber Catholics.

Slovakia

Slovakia is smaller in both size and population than the Czech Republic and somewhat less developed, although it, too, is successfully making the transition from a centrally planned economy to a free-market economy. Slovakia is a land of rough mountains, forests, and high pastures.

Slovakia has a significant ethnic Hungarian minority, whose status has caused some tension with Hungary. Although Slovak is the official language, Hungarian is also spoken in many areas. The capital is Bratislava, with about 450,000 inhabitants. Roman Catholicism is the principal religion, although significant minority religions include Protestantism, Eastern Orthodoxy, and Judaism.

Hungary

Having sided with Germany in World War II, Hungary suffered a particularly harsh occupation at the hands of the victorious Soviets. The Hungarian uprising of 1956 was brutally put down by Moscow. In 1968, Hungary began a process of liberalizing its economy, in what came to be known as "goulash communism," after the country's trademark spicy stew.

Hungary is a land of vast fertile plains, and prior to World War II its economy was primarily agricultural. Post-war industrialization was badly managed by the government and large portions of the country's heavy industries collapsed after the end of communist rule in the early 1990s. Although the initial period of transition between a controlled and a free-market economy was difficult, Hungary has weathered the crisis and today enjoys the third highest per capita GDP in the region, after Slovenia and the Czech Republic. Hungary's capital is Budapest (population 1.8 million).

Hungarians are descendants of an ancient Asian tribe called the Magyars. Almost 70 percent of Hungary's people are Roman Catholic, while Protestants account for another 25 percent.

Slovenia

A major Eastern European success story, Slovenia is the first former Yugoslav country to become a member of the European Union. Its economy is the strongest in the region, with a similar per capita GDP as Portugal. Slovenia's historic ties to the West explain, in part, its successful transformation into a Western-style capitalist democracy. After the collapse of the Austro-Hungarian Empire in 1918, the Slovenes opted to join with the Serbs and Croats to form what would become known as Yugoslavia. After a brief 10-day war in 1991, Slovenia became the first part of the unraveling Yugoslavia to become independent.

About the size of Massachusetts, Slovenia is a largely mountainous country, about half of which is forested. Its population is predominately Slovene, with small Serb and Croat minorities. The overwhelming majority of Slovenes speak Slovenian, a South Slavic language closely related to Serbian and Croatian, which most Slovenes also speak. About 70 percent of the population is Roman Catholic. The political and cultural heart of the Slovene nation is the capital, Ljubljana, a charming and highly cultured city of some 264,000 people.

Romania

Both Romania and Bulgaria joined the European Union in January 2007, but both have formidable economic and political hurdles to overcome to be a free-market economy. Romania inherited from the communist era a largely outmoded industrial base and a huge agricultural sector that accounts for a whopping 40 percent of its workforce. Although Romania has enjoyed strong growth in recent years and has considerable mineral resources and oil and gas reserves, it continues to be plagued by poverty and corruption.

Romania's center is dominated by the rugged Carpathian Mountains that run north-south and the Transylvanian Alps that run east-west. West and north of these ranges lies Transylvania. The capital of Bucharest is by far Romania's largest city, with more than 2 million people. Unfortunately, the once-beautiful capital was severely disfigured by the systematization building programs of the dictator Nicolae Ceausescu during the 1970s and '80s, which destroyed more than 20 percent of old Bucharest. The population is almost 90 percent Romanian, but includes a sizable Hungarian minority and a smaller Gypsy minority. Unique in Eastern Europe, Romanian is a Romance language of the Italic group, closely related to Italian, Spanish, and Portuguese. The main religion in Romania is Romanian Orthodox.

Bulgaria

Bulgaria is an ancient land that for much of its modern history (from the late fourteenth century to 1878) was under Turkish rule. In the twentieth century, it found itself under Soviet domination from 1945 to 1990. Although Bulgaria has experienced stability and economic growth since 1996 and joined the European Union in early 2007, it still has to cope with high rates of poverty and unemployment.

Bulgaria is a mountainous country bisected by the Balkan Mountains. It is one of three Eastern European countries to have a Black Sea coastline (the other two are Romania and Ukraine). Bulgaria has a sizable Turkish minority (9.5 percent) and a smaller Gypsy minority. Bulgarian, a Slavic tongue, is the language of the land, with Turkish spoken in Turkish enclaves. As is true of most Eastern European capital cities, Sofia (population 1.1 million) is several times larger than Bulgaria's next largest city.

Out on Their Own: Belarus, Ukraine, and Moldova

The remaining three Eastern European nations are the former Soviet republics of Belarus, Ukraine, and Moldova and have lagged behind the "EU-ready" nations in the region.

Belarus

Sandwiched between the Baltic states to the north and Ukraine to the south is the former Soviet Socialist Republic of Belarus. Although it declared its independence in 1991, Belarus maintains closer cultural, economic, and even political ties with Russia than any of the former Soviet socialist republics (SSRs) or satellite nations. In fact, Belarus and Russia signed a treaty in 1999 that set the framework for increased political and economic integration. Implementation of the treaty, however, has been slow in coming.

The economy of Belarus remains closely linked to Russia's and isolated from the West. Unlike its Baltic neighbors, Belarus has resisted free-market reforms in favor of "market socialism." The result has been increased government intervention in the private sector and high rates of inflation and poverty.

Landlocked Belarus is low-lying, with many rivers and lakes, and lots of swamps (its peat marshes are an important, if dirty, energy resource). About one third of the land is forested. The country's main population and industrial center is its capital, Mensk (or Minsk), with nearly 1.7 million residents. The population of Belarus is predominantly Belarusian, but includes a large Russian minority (more than 11 percent). The people speak Belarusian (White Russian), a Slavic language closely related to Russian. Eastern Orthodoxy is the primary religion.

Ukraine

Ancient Ukraine was the center of the first Slavic state, Kievan Rus, which later moved to Moscow. Over the centuries the Ukrainians developed a unique but related culture with their Moscow-centered relatives the Russians.

With abundant natural resources and vast expanses of fertile soil, Ukraine was one of the chief industrial engines of the old Soviet Union as well as its primary breadbasket. Ukraine today remains overly dependent on Russia as a trading partner, and its major industries are still mostly state-controlled. Ukraine's inability to put the Soviet past behind it and embrace free-market reforms, coupled with corruption and technical instability, are serious obstacles to its economic development.

Eye on the Environment
An especially troubling legacy of the Soviet era is the nuclear power plants on which Ukraine depends for about 43 percent of its electricity. Radiation from the Chernobyl nuclear power plant, which exploded with such devastating consequences in 1986, is still contaminating Northeastern Ukraine and parts of Belarus.

With an area of 233,000 square miles, Ukraine is the largest country entirely in Europe. (Russia is larger, but it straddles Europe and Asia.) Nearly 78 percent of Ukraine's population is ethnic Ukrainian, and about 17 percent is Russian. The language of the country is Ukrainian, a Slavic tongue similar to Russian and Belarusian. The vast majority of Ukrainians are Eastern Orthodox; about 13.5 percent practice Ukrainian Catholicism. The capital and largest city is Kyiv (or Kiev), with more than 2.6 million people.

Moldova

Even after Moldova became independent in 1991, Russia maintained a presence on the Eastern fringe of the country, east of the Dniester River, to support the majority Russian and Ukrainian population there.

Moldova today is one of the poorest countries in Europe and in 2001 had the dubious distinction of being the first former Soviet Republic to elect a communist president. Its dependence on the former Soviet republics for its raw materials and trading relations is a great economic vulnerability. Despite its many problems, Moldova has privatized its farmland (with U.S. help) and has been showing steady growth in the early 2000s.

Moldova is a small but densely populated country. The capital and largest city is Chisinau, with 750,000 inhabitants. Moldovan/Romanians are the majority population, although Slavs, mostly Russians and Ukrainians, are the majority east of the Dniester River (in the disputed Transnistria region). Moldovan (almost the same as Romanian) and Russian are the official languages. Moldova's people are largely Romanian Orthodox in their faith.

Ethnic Powder Kegs: The Former Yugoslavia and Albania

The third subregion of Eastern Europe encompasses five countries on the western side of the Balkan Peninsula, as shown on the map of Eastern Europe. (Although Slovenia is also a former Yugoslav country, it has been grouped in this chapter with the more stable and prosperous Eastern European nations that are the newest members of the European Union.) The fractured, mountainous landscape reflects the ethnic splintering that has long dominated the history of this subregion.

Fractures and Feuds: The Former Yugoslavia

Yugoslavia was patched together after World War I out of an assortment of disparate and often conflicting states, territories, and ethnic and religious groups. The Nazi occupation of the country during World War II was particularly harsh and bitterly fought by opposing partisan guerilla armies—one monarchist, the other communist under the ironfisted leadership of Marshall Tito. After the war, Tito imposed a unique form of communism on Yugoslavia that combined rigid political control with a certain amount of economic and cultural flexibility. In 1948, Tito broke with the Soviet Union and went his own way. Yugoslavia never became a Soviet satellite or a member of the Warsaw Pact.

Tito died in 1980, and by the early 1990s, with Soviet communism about to collapse, Yugoslavia began to fall apart. Slovenia, Croatia, and Macedonia were the first to secede in 1991. Bosnia and Herzegovina followed suit in 1992. Yugoslavia, consisting of Serbia and Montenegro, reconstituted itself as the Federal Republic of Yugoslavia. Horrendous atrocities were committed on all sides during the following period of civil war. Serbia, in particular, attempted to expand and control more territory and the term "ethnic cleansing" became associated with the region. NATO intervened militarily in 1999 and today an uneasy peace holds, with the help of UN observer forces in place to monitor the various factions.

Today Slovenia, Croatia, Bosnia-Herzegovina, and Macedonia are all fully independent countries.

The Former Yugoslav Nations: An Overview

Geography, history, language, and religion have all played a role in the *balkanization* of the Balkans. The entire region is dominated by the Dinaric Alps, whose rugged mountains and isolated valleys provide the framework for the development of individual cultures in isolation. Because the mountainous divides hindered cultural exchange, groups tended to develop separately. The result is today's mix of neighboring, but often very distant, ethnic groups—Serbs, Croats, Slovenes, Slovaks, Hungarians, Albanians, Bulgarians, Montenegrins, Macedonians, and Muslims.

def•i•ni•tion

The term **balkanization** has come to mean the breaking up of a country or region into smaller, hostile units, which certainly describes the history of the Balkan Peninsula.

The most common languages in this subregion are Serbian, Croatian, Bosnian, Slovenian, and Macedonian. Religion has also been an extremely divisive issue in the former Yugoslavia, with Eastern Orthodox Serbs pitted against Roman Catholic Croatians or Muslim Bosnians. Here's a rundown of the new countries that emerged from the dissolution of Yugoslavia:

- **Bosnia and Herzegovina:** Capital: Sarajevo (Mostar is the traditional Herzegovinian capital). Ethnic groups: Bosniak (44 percent), Serb (31 percent), and Croat (17 percent). Languages: Bosnian, Serbian, and Croatian. Religions: Islam (48 percent), Orthodox (37 percent), and Roman Catholic (14 percent).

- **Croatia:** Capital: Zagreb (population: 870,000). Ethnic groups: Croat (90 percent), Serb (4.5 percent), with small Bosniak, Slovenian, and Hungarian minorities. Languages: Croatian and Serbian. Religions: Roman Catholic, Orthodox, and Islam.

- **Macedonia:** Capital: Skopje (population: 430,000). Ethnic groups: Mostly Macedonians with a sizable Albanian minority and some Turks, Gypsies (Roma), and Serbs. Languages: Macedonian and Albanian are official languages; Turkish is also spoken. Religions: Eastern Orthodox and Islam.

- **Serbia:** Capital: Belgrade (with 1.6 million residents, this is the largest city in the entire area). Ethnic groups: Mostly Serb (more than 82 percent), with a large Albanian minority in Kosovo (about 16 percent). Languages: mostly Serbian, some Albanian. Religions: Serb Orthodox and Islam.

- **Montenegro:** Capital: Podgorica. Split from Serbia on May 21, 2006, to become independent. The 650,000 population of the country is pretty divided amongst its dozen or so larger cities. The people are mostly Montenegrin, a branch of the Serbian people, and they speak the Montenegrin dialect of the general Slavic tongue common to the Balkans. Most people are Montenegrin Orthodox, although there are small Catholic and Moslem minorities.

◆ **Slovenia:** Detailed earlier in this chapter.

Terra-Trivia

Despite regional and ethnic difference, the predominant language throughout the former Yugoslavia used to be called Serbo-Croatian. No longer. It is now called either Serbian or Croatian or Bosnian, depending on the speaker's politics and ethnicity. There's not even a shared alphabet. Croatians and Slovenes use the Latin alphabet, while Serbs, Bosnians, Montenegrins, and Macedonians use the Cyrillic alphabet.

Albania: Chinese Ties and Isolationism

The last of the Eastern European countries described in this chapter is also one of the poorest but most rapidly transforming in Europe. Albania emerged from more than 400 years of Turkish rule in 1912 only to be conquered by Italy, and then fall under a brutal Stalinist communist regime after World War II. In 1961, Albania broke with Moscow and aligned with Maoist China. After Mao's death in 1978, Albania completely isolated itself from the rest of the world and sank into economic ruin. The end of Albania's communist regime in 1992 initiated a difficult decade of transition and adjustment, marked by crime, corruption, high unemployment, and involvement in its neighbors' conflicts, especially in Yugoslavia. Per capita GDP expanded a dramatic 85 percent between 1990 and 2000. Economic expansion has continued to expand since 2000.

Albania is a mostly mountainous country, with most of its farmlands and population concentrated in the western coastal lowlands. The country is heavily dependent on agriculture, which accounts for about 50 percent of its GDP, and on money sent home by Albanians working abroad, mostly in Italy and Greece. Both Albanian and Greek are widely spoken, and Islam (70 percent) and Albanian Orthodoxy (20 percent) are the main religions. The capital, Tirana, has a population of nearly 600,000 (though some estimates place it much higher).

The Landforms of Eastern Europe

Large expanses of plains dominate Eastern Europe's northern and eastern sections. Although the area's flat plains have historically invited hostile armies to march across them with depressing ease and frequency, they have also been a blessing. With no intervening mountain range, the warmer air from the Atlantic Ocean and Baltic Sea helps to moderate the climate. Although still quite cold, temperatures in the plains are warmer than might be expected for their latitude. The fertility of the plains is also responsible for the region's agricultural abundance.

Eastern Europe—physical features.

The Mountains

Moving south and west through Eastern Europe, the plains give way to mountains. The Carpathian Mountains and Transylvanian Alps are located in the center of the region, the Balkan Mountains are in the southeast, and the long chain of the Dinaric Alps are in the southwest. These mountains and their many valleys and basins have helped nurture and separate the region's diverse cultures.

The Great Waters

Eastern Europe is sandwiched between three large bodies of water: to the north, it's bounded by the Baltic Sea; to the southeast, the Black Sea; and to the southwest, the Adriatic Sea. All except six of the region's countries have a coastline on one of these three seas. (Belarus, the Czech Republic, Slovakia, Hungary, Moldova, and Macedonia are all landlocked.)

The region is also laced by several important rivers. The northern plains are drained by Poland's Vistula River, which empties into the Baltic Sea. In the east, waters from the vast plains of Belarus and Ukraine drain south through the Dnieper River to the Black Sea. The region's most important river, the Danube, winds its way through most of the countries in the south to exit in the Black Sea. It runs through or touches more countries than any other river on Earth.

In the News: The Soviet Environmental Legacy

The defunct Soviet Union left its former republics and satellites with a dark double legacy: shattered economies and a blighted environment. During the Soviet decades, the region's state-run economies were focused on increasing industrial output no matter what the cost to the environment or to the health of its citizens. Aging, sub-standard, and largely unregulated factories, mines, and power plants spewed toxic fumes into the air and dumped pollutants into the water supply, with dire consequences for the health of humans, wildlife, and forests. The deadliest symbol of the Soviet Union's failed industrial and environmental policies is the Chernobyl nuclear power plant, whose shuttered ruins are still a threat to the people of Ukraine and Belarus. Cleaning up is proving to be a costly job.

The Least You Need to Know

- ◆ This region's 19 countries include former Soviet republics, former satellites, and new offshoots from the former Yugoslavia.

- ◆ All of the countries in this region are undergoing a transition from communist economies to Western-style free-market economies, some more successfully than others.

- ◆ The most successful economies of Eastern Europe—Estonia, Latvia, Lithuania, Poland, the Czech Republic, Slovakia, Hungary, Slovenia, Romania, and Bulgaria—are the newest members of the European Union.

- ◆ The post-communist breakup of Yugoslavia was accompanied by ethnic violence, civil war, ethnic cleansing, and foreign intervention.

- ◆ One of Europe's great rivers, the Danube, flows through the region's southern tier before emptying into the Black Sea.

Chapter 16

Russia in Transition

In This Chapter

- ◆ Tracking the tsars: a look at Russian history
- ◆ Assessing the new Russia—without the republics
- ◆ Touring the tundra and taiga
- ◆ Meeting the Russian people
- ◆ Looking toward the future

It's been more than a decade and a half since the disintegration of the Soviet Union in 1991 and the reemergence of Russia as a separate country. Yet there is still good reason to call this region the "new" Russia, for Russia is still very much in the process of defining itself and its new place in the world. It may have lost its empire, but Russia is still the largest country in the world.

Communists and Free Markets

In the Bolshevik Revolution of 1917, the tsar and his family were murdered by the Bolsheviks, bringing to an end 300 years of Romanoff rule. The communists took control after a violent civil war and established the Union of Soviet Socialist Republics, or USSR, in 1922.

Russia.

Although the Nazis attempted to conquer Russia in World War II, the Red Army was able to turn them back at the famous Battle of Stalingrad. The victorious Russians pushed the Nazis back and began grabbing territory and establishing satellite countries, including the entire eastern portion of Germany.

During the cold war years, the Soviet Union engaged in an economically devastating arms race with the West (specifically, the United States). Both sides stockpiled nuclear weapons. The communists slowly lost control of their empire and on December 21, 1991, the USSR was replaced by the new Commonwealth of Independent States, an economic and security union currently consisting of Russia, now officially known as the Russian Federation, and 11 of the newly independent former Soviet republics.

The Loss of the Former Republics

Just how geographically different is Russia from the Soviet Union of yesterday? This section will help you compare the old and the new. The Union of Soviet Socialist Republics consisted of 14 republics and 1 Soviet Federated Socialist Republic (Russia):

Armenia	Lithuania
Azerbaijan	Moldova
Belarus	Russia
Estonia	Tajikistan
Georgia	Turkmenistan
Kazakstan	Ukraine
Kyrgyzstan	Uzbekistan
Latvia	

Terra-Trivia

Until 1867, the area that is now Alaska was Russian territory. In a deal driven by then-Secretary of State William Seward, the United States purchased Alaska for just $7.2 million. Although critics of the purchase dubbed the new territory "Seward's Folly," Alaska's gold and oil alone have earned back its purchase price countless times over.

Today, all 15 former Soviet republics are independent nations and members of the United Nations.

Before the breakup, the USSR's land area of 8,649,489 square miles made it the largest country in the world. On its own, Russia is still the world's largest country, with a land area of 6,592,800 square miles—more than 11 percent of the earth's total land surface. Nonetheless, Russia lost more than 2 million square miles, or almost a quarter of its former land area.

The losses in terms of population were even more dramatic. The population of the former Soviet Union was about 289,000,000. Russia's population is now 142,800,000, a 50-percent drop. It's interesting that Russia's own population has declined in recent years. With a birthrate around 10 per 1,000 and a death rate around 14 per 1,000, Russia is experiencing negative population growth, as are many of the former Soviet republics and satellites. Ethnic Russians represented only about 52 percent of the population of the former USSR. Today, they constitute nearly 82 percent of the population of post-Soviet Russia.

Despite the loss of its East European republics, Russia still spans the same 11 time zones that the Soviet Union did. The reason for this is that after the breakup of the Soviet Union, Russia retained a small, fragmented area, called the Kaliningrad Oblast, squeezed between Poland and Lithuania on the Baltic Sea, in what was once part of East Prussia. Kaliningrad adds an extra time zone to Russia's 10.

In terms of population, the USSR was the world's third-largest country (behind China and India). Russia today ranks eighth in the world in population (behind China, India, the United States, Indonesia, Brazil, Pakistan, and Bangladesh).

An Unforgiving Climate

Almost all of Russia is located above 50 degrees north latitude, roughly equivalent to Canada's latitude. Saint Petersburg is located at about 60 degrees north latitude, placing it less than 7 degrees south of the Arctic Circle and about 2½ degrees north of Alaska's capital, Juneau.

Although Russia's latitude parallels Canada's, Russia is colder. Northern Canada encompasses the huge Hudson Bay, which has a moderating effect on climate. The fact that Arctic Canada consists mainly of islands and waterways also helps to temper the climate somewhat. Russia, on the other hand, occupies the extreme north of the world's largest uninterrupted continental landmass, Eurasia, with no large, interspersed body of water to moderate the harsh temperatures.

During the northern Siberian winter, temperatures plummet and the sun never rises above the horizon for months on end. High pressure is indicative of cold, dry air. With no major body of water to moderate temperatures, Siberia is subject to tremendous ranges in annual temperature. In parts of central Siberia, the normal January temperature of less than –50 degrees F rises to an average of 50 degrees F in July—an astonishing 100-degree range.

Because the temperatures are so low, the air can hold little moisture. The result is precipitation totals of less than five inches between November 1 and April 30. Although the summer months bring slightly more precipitation, especially in the western parts of Russia, the bulk of Russia is quite dry. As though cold and dry weren't bad enough, much of Russia's northern lands are covered with permafrost.

Permafrost, or permanently frozen subsoil, poses all sorts of problems to permanent structures and transportation lines. If it remains frozen, it provides a hard surface. Beware, however, if it thaws. Road-building equipment, houses, and chunks of road surface have simply disappeared into the quicksand-like ooze of thawed permafrost. Rail lines built across permafrost and warmed from the friction of train traffic or solar radiation look like they have been strung across a giant washboard.

Feeding its people has always been a problem for Russia. Even the USSR, with the more fertile lands of Belarus, Ukraine, and Moldova, had trouble producing enough food. The scenario in Russia is worse now that these lands are independent and no longer part of the country.

By Land or by Sea

Vast mountains, vast plains, vast forests—Russia is sort of a vast-fest. It also has some of the world's largest lakes and rivers.

The Land—and Lots of It!

In the west, Russia's European territory is dominated by the Russian Plain, an extension of the Northern European Plain. The northern part of the Russian Plain is primarily woodland, while the southern reaches contain Russia's primary agricultural lands. To the south, the plains terminate between the Black and Caspian seas in the Caucasus Mountains. The eastern edge of the plain (and the end of continental Europe) is marked by the long north-south chain of the Ural Mountains. The Ural chain is a primary source of the fuels and minerals that supply Russia's industry. The Ural chain also marks the traditional dividing line between the continents of Europe and Asia.

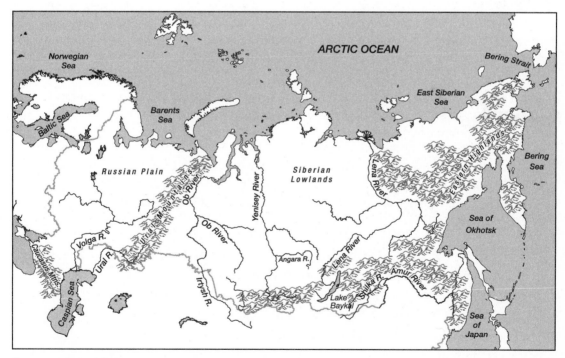

Russia—physical features.

Beyond the Urals lies the often frozen expanse of Siberia. The first thousand miles or so consist of the Siberian Lowlands. They're dominated by needleleaf evergreens and tens of thousands of square miles of swamps. East of the lowlands, the terrain begins to rise. First comes the Central Siberian Plateau, and, on Russia's Pacific coast, the mountainous Eastern Highlands. Much of the plateau and highlands are covered with needleleaf deciduous trees called *taiga*.

As inhospitable as this land might seem, it contains great wealth. Beneath the frozen surface lie tremendous energy reserves. Russia has more than 16 percent of the world's known coal reserves (second to the United States, with 23 percent). It's a world leader in uranium reserves (second only to Australia), and ranks sixth (or seventh, depending on who's counting) in the world in proven petroleum reserves. Perhaps the most staggering statistic is that Russia has more than one third of the world's total known natural gas reserves.

Frozen Waters

Russia is bounded by several large bodies of water. Starting in the north and going clockwise from there are the Barents Sea, the frigid Arctic Ocean, and the East Siberian Sea. To the northeast, Russia is separated from the United States (Alaska) by the Bering Strait. In the east are the Bering Sea, the Sea of Okhotsk, the Sea of Japan, and the Pacific Ocean. To the south there is mostly land, but Russia also has a southern coastline on the Aral Sea, the Caspian Sea, and the Black Sea. In the west, Russia has access to the Baltic Sea via the fragmented bit of territory called Kaliningrad Oblast, as well as a coastline on the Gulf of Finland.

Russia borders on the Caspian Sea, which is the world's largest lake in terms of surface area. (Because the Caspian is landlocked, it's considered a lake, even though it's called a sea and contains saltwater.) Lake Baykal, known for its cold, crystal-clear water, is the earth's sixth-largest natural freshwater lake (in surface area). It is also by far the earth's deepest lake, with a maximum depth of 5,316 feet (more than a mile deep!).

Russia also has its share of major rivers, but the most important of them by far is the Volga. Europe's principal river in terms of length, volume of water, and size of drainage basin, the mighty Volga and its tributaries flow through the heart of the populous Russian Plain before finally draining into the Caspian Sea. But although the Volga might be a big deal in Europe, it is outranked in terms of length, volume of water, and size of drainage basin by four of Russia's Asian rivers. The Volga is 2,194 miles long. Going from west to east, the four longer rivers in Russia are the Ob'-Irtysh (3,362 miles), the Yenisey-Angara (3,449 miles), the Lena (2,734 miles), and the Amur-Shilka (2,744 miles).

People of the Steppes

To understand the distribution of Russia's population, think west and think south. The overwhelming mass of the Russian population lives west of the Ural Mountains on the great Russian Plain, which is also the country's agricultural heartland. The population in this area is concentrated mostly in the southeast and gradually thins as you move to the north. East of the Urals, the huge expanse of Siberia is largely uninhabited. The hardy souls who do subsist in this frozen realm are clustered primarily along its southern reaches.

Geographically Speaking

Though Russia has 11 cities with populations of more than 1 million, you have to remember only two of them: the capital, Moscow (now Europe's largest city with 14.5 million people), and the former capital, Saint Petersburg (with 5.5 million people). In addition to being Russia's largest cities, they are also the twin centers of its politics, history, and culture.

Russia's population is more than 80 percent ethnic Russian; the remaining 20 percent includes a variety of other ethnic groups including Tatars and other Turkic peoples such as Chuvash and Bashkri, Ukrainians, Mongolians, Caucasians, Moldovans, and Finno-Ugrians. Although Russian is the official and almost universal language of the land, many citizens who are not ethnic Russians also speak their native language.

Since the breakup of the Soviet Union, Russia has experienced a tremendous religious resurgence. Most of the growth has occurred in the country's ancient and predominant religion, the Russian Orthodox Church.

Terra-Trivia

Religion in Russia was suppressed during the communist years. The official "religion" of the state was atheism, and many Russians still claim no religious affiliation. Although the predominantly Muslim Central Asian republics of the old USSR are now independent, Russia still has a large Muslim minority, about 19 percent of the population.

Russia in the Twenty-First Century

Russia today is a country of enormous potential and equally enormous problems. It is a major industrial power, but its factories and plants are outmoded, inefficient, and often decrepit. Russia is immensely rich in minerals and fossil fuels, but extracting them from remote, inhospitable areas can be difficult and costly. In addition, Russia's economy is now overly dependent on exported raw materials, making it highly vulnerable to fluctuations in market prices. In today's truncated Russia, less than 8 percent of the land is arable and the agricultural sector remains in crisis mode, plagued by a crumbling rural infrastructure and farm privatization that has been slow, difficult, and controversial. Despite the many problems, however, Russian grain production is rising and the terrible Soviet-era food shortages seem to be a thing of the past.

Unlike its former satellites in Eastern Europe, Russia was slow to implement free-market reforms and as a result saw its economy contract for five years after independence. After a brief period of growth in 1997, Russia was hard hit by the world financial crisis of 1998, which forced the government to devalue the ruble (the Russian currency) and default on its debts. But today, in the early years of the twenty-first century, the Russian economy is growing once again. Still, the new Russia has many serious hurdles to overcome before it completes its transition to a free-market economy. The long list includes pervasive corruption, organized crime, high levels of poverty, shaky banks, a less than independent judicial system, and an environment scarred by decades of unregulated industrialization and neglect.

Perhaps one of the greatest challenges facing Russia today is defining its role in the new post-cold war world order. Russia is no longer the superpower it was just over a decade ago. Even its space program, the centerpiece of Soviet science and technology, is

Geographically Speaking

At the end of World War II, the Soviets grabbed Japan's Kuril Islands. Japan wants the islands back; Russia doesn't want to give them up. This lingering political impasse has kept the two nations from, believe it or not, signing the peace treaty ending World War II and forging an economic alliance that would be so beneficial to both parties.

teetering on the brink of collapse for lack of funds. This loss of power and prestige does not sit well with the Russian people. How they cope with their transition and go about fulfilling their country's vast potential will be a matter of concern over the coming decades not only for Russia and its neighbors, but for the entire world.

In the News: Chechnya

For much of the 1990s and into this century, the name Chechnya has been a fixture on newspaper front pages around the world. During the Soviet era, Chechnya was part of the mostly Muslim Chechen-Ingush Autonomous Soviet Socialist Republic located in the Caucasus Mountains. When the Soviet Union collapsed in 1991, Chechnya declared itself independent, a move Russia refused to recognize. Tensions escalated into violence, Russian troops were sent to Chechnya, and a bloody war broke out in 1994. Two years and an estimated 70,000 casualties later, a ceasefire was declared and Soviet troops pulled out from what was now an essentially autonomous Chechnya. Although this was perceived as a victory for Chechnya and a humiliation for the Russian army, the war resolved nothing.

In 1999, with Chechnya falling into anarchy, the Russian-Chechen conflict took on a new and more dangerous turn. Russian troops attempted to impose order in the area. The Moslem Chechens responded by taking the war into the Russian heartland with terrorist bombings, even in Russia's capital of Moscow. The Chechen issue remains unresolved.

The Least You Need to Know

- Russia's emperors were called tsars. The last tsar was overthrown in the communist revolution of 1917.

- The USSR's 15 former republics have all spun off and become separate countries (including Russia).

- Russia is the world's largest country in area and eighth largest in population; Moscow and Saint Petersburg are its largest and most important cities.

- Its high latitudes give Russia an extremely cold climate.

- Most of the region's people speak Russian and are Russian Orthodox.

- Russia is in the process of making a slow and difficult transition from a rigid state-controlled economy to a free-market economy.

Chapter 17

China: The Awakened Giant

In This Chapter

- ◆ The immensity of China and its history
- ◆ China today: calculating the pluses and minuses
- ◆ A return to forgotten Mongolia
- ◆ North Korea: East Asian flashpoint

In many ways, this huge and hugely important country is still a developing nation. In terms of poverty and unemployment rates, living standards, environmental conditions, and percentage of the work force involved in agriculture, China has a long way to go to catch up with the developed nations of the world.

On the other hand, China's gross domestic product has now surpassed that of Japan (in fact, it is twice that of Japan), making it the world's second largest economy after the United States. Like the other transition countries covered in this part of the book, China is undergoing a major transformation from a rigidly state-controlled economy to a more open, Western-style free-market economy. The difference between China and the former Soviet republics and satellites is that China politically, if not economically, is still a rigidly controlled communist state.

China is the primary economic and political power in the area, but Mongolia, China's remote and physically isolated neighbor to the north; and North Korea, China's politically isolated and increasingly dangerous communist neighbor to the northeast, are also players in the region.

Asia's Megastate: The People's Republic of China

China is huge. The world's third-largest country in area (behind only Russia and Canada), China occupies 3.7 million square miles of Eastern Asia. It has more than 3,400 offshore islands, Hainan being the largest of them. With more than 1.3 billion people, China is the world's most populous country. Fifty of its cities have more than a million people apiece; nearly one out of every five people on Earth is Chinese.

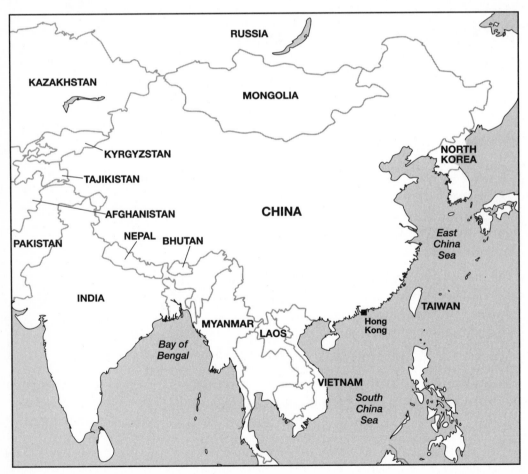

China and its neighbors.

China's neighbors to the north are Mongolia and Russia; to the northeast is North Korea; to the east are the Yellow and East China seas; to the south is the South China Sea, Vietnam, Laos, Myanmar, India, Bhutan, and Nepal; to the west, China touches Pakistan, Afghanistan, and Tajikistan; and to the northwest are Kyrgyzstan and Kazakhstan.

China is so extensive that it includes almost every type of climate, every degree of rainfall, and every topographical feature imaginable. Its surface is 43 percent mountains, 26 percent plateaus, 19 percent basins, and 12 percent plains.

In the Beginning: North-Central China

The ancient cradle of Chinese civilization is the 3,395-mile Huang He (Yellow River) basin. In its lower reaches, the Huang River flows through the fertile and densely populated North China Plain, one of the country's most important agricultural regions. Located in the northeastern corner of the plain are China's capital, Beijing, home to some 8.6 million people, and the important industrial city and commercial port of Tianjin, with another 4.7 million people. Surrounding these two megacities is a major industrial and oil-drilling area.

China's Furnace: The Southern Section

South of the Northern Plain is a varied landscape of mountains, plateaus, hills, and river basins, dominated by the 3,900-mile-long Chiang Jiang (Yangtze River), China's (and Asia's) longest river and the third longest in the world. The Chian Jiang basin is China's major rice- and tea-growing region and home to about a third of its population. The Chiang Jiang drains into the East China Sea near Shanghai, China's most populous city, home to about 10 million people in the city proper and more than 12 million in its metropolitan area. Shanghai is also China's largest port and a commercial and financial center now rivaling Hong Kong in importance.

Upriver from Shanghai is the city of Nanjing. Nearby is one of the world's most outstanding engineering projects: the Grand Canal linking Beijing to the south. The busy artificial waterway is the oldest and longest canal in the world.

China's deep south is known as the "furnace" for its extreme heat and humidity. The Zhu Jiang (Pearl River) basin area is tropical in all respects: heavy rainfall, high temperatures, and lush vegetation. Two, sometimes even three, rice crops are harvested in a year here. The Zhu Jiang flows through Guangdong Province, birthplace of

China's modern-day economic miracle, before draining into the South China Sea. Located in the mighty Zhu Jiang delta are Hong Kong, the economic powerhouse and former British colony now restored to China; Macao, the former Portuguese colony now also part of China once again; and Shenzhen's Special Economic Zone, one of China's first experiments in modernizing its economy. At the north end of the Zhu Jiang delta is the great trading metropolis of Guangzhou (Canton). Capital of Guangdong Province and China's southern gateway, Guangzhou has a population of about 6.6 million people. China's largest island and smallest province, Hainan, lies just off Guangdong Province's southwestern Leizhou Peninsula.

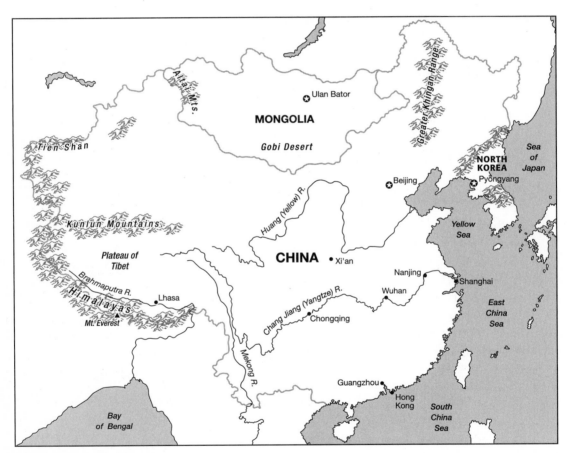

China—physical features.

The Top Half: Northeastern and Northwestern China

Northeastern China is half of historic Manchuria (which used to include Mongolia). This extensive area, consisting of a large central plain bordered on the east and west by mountain chains, is China's "Siberia": it's cold and has a short growing season.

Northwestern China comprises three large desert basins, each surrounded by great mountain chains. The entire area is sparsely settled. Aside from the Silk Route, not much else has passed through this area.

Terra-Trivia _____

The Great Wall of China, early portions built by the great emperor of the Qin Dynasty, Shih Huang Di (221–207 B.C.E.), was built at tremendous cost in materials and lives. The 3,000-mile wall was supposed to keep out northern invaders, but proved ineffective a number of times, especially during the great Mongol invasion.

Ancient Tibet: China's Reluctant Southwest

Tibet (or Xizang, its Chinese name), the largest, least populated (2.2 million people), and most isolated of China's areas (it's actually an autonomous region, not a province), was conquered by the Chinese in 1950. The Tibetan capital is Lhasa, renowned for the Potala Palace, the traditional home of the Dalai Lama. Tibet is located in Southwestern China, just northeast of India. It is called the "roof of the world" and with good reason: with an average elevation of more than 16,000 feet, it is the highest region on Earth. Tibet's central plateau (also the world's highest with an average elevation of 15,000 feet) is surrounded by mountains to the north and west and by the mighty Himalayan Mountains to the south. Tibet's share of the Himalayas includes Mount Everest, the world's highest peak at 29,028 feet, which straddles the Tibet-India border.

Terra-Trivia _____

The Dalai Lama, the spiritual and temporal leader of Tibet and Tibetan Buddhists across the world, fled Tibet in 1959, after a failed uprising against the Chinese occupiers, when he was in his mid-20s. He now lives in exile in India but lectures all over the world in nonviolent opposition to Chinese rule in Tibet (for which he received the 1989 Nobel Prize for peace).

Tibet's plateau is Asia's principal watershed. It's the source of the Huang (Yellow), Chang Jiang (Yangtze), Mekong, Ganges, Indus, and Brahmaputra rivers. Tibet is semiarid (it receives only 15 inches of precipitation annually) and is generally cold and windy. Its people are mostly Tibetan (although a sizable Chinese minority now lives there) and speak Tibetan. They are strongly Lamaist (Tibetan Buddhist), although China encourages atheism. Most Tibetans survive through herding and subsistence farming.

The Birth of Modern China

After the last emperor was forced to abdicate in 1912, various "democratic" groups and warlords fought for dominance. In 1928 a Nationalist, democracy government gained control. Civil strife became the norm through the 1930s as radical communist groups challenged the democratic central government. Civil war broke out but was interrupted by the Japanese occupation of parts of China starting in 1931.

With the defeat of Japan at the end of World War II, however, the civil war resumed, but this time the communists, under Mao Zedong, took control of China. In 1949, Mao proclaimed the new People's Republic of China. The nationalists, under Chiang Kai-shek, withdrew to the island of Taiwan.

The People's Republic of China: The First Half-Century

Under Mao, China's communist government strictly controlled all aspects of its citizens' lives. Religious expression was suppressed and a cult of personality developed around Mao. Millions died in misguided attempts at reengineering society, such as the "Great Leap Forward" of 1958 and especially the Cultural Revolution of 1966 to 1969. After Mao, various communist leaders tried to continue his policies but gradually opened up to some market reforms.

The Economic Miracle

Since introducing market reforms in 1978, China's gross domestic product (GDP) has quadrupled and is now second only to that of the United States. China's enormous population, however, means that its per capita GDP is considerably less robust. In 2002, it was on the same level as Belize and just above Suriname. Still, China's potential as a producer of goods and services and as a market for the rest of the world's goods and services has barely been tapped.

Over the past several years China's economic growth rates have been in the extraordinary 7 to 8 percent range. Most impressive is the fact that growth has been in both the industrial and the agricultural sectors, the latter being of particular importance to a country historically prone to devastating famines. China's admission to the World Trade Organization in 2001 was not only an acknowledgement of its economic reforms, but also a boost to its growth and an incentive to continue liberalizing not only its economy, but its government and legal system as well.

On the energy front, China has vast coal deposits and is well endowed with petroleum, natural gas, and oil shale. It also has the potential to be the world's biggest producer of hydroelectric power. China is known to have reserves of 153 different minerals and many of them in abundant quantities. Its supplies of iron, copper, aluminum, tin, zinc, lead, and others are among the biggest in the world.

Eye on the Environment

China's fabled Chiang Jang, better known in the West as the Yangtze River, is both a national lifeline and a flood-prone killer. More than a million people are estimated to have died in Yangtze floods over the past 100 years. Now, in what has been called the nation's largest public works project since the Great Wall, China proposes to tame the mighty Yangtze with what will be, when it's completed in 2009, the world's largest hydroelectric dam, the controversial Three Gorges Dam.

The dam's 26 turbines are expected to generate one ninth of China's electrical power. The reservoir is expected to flood nearly 400 square miles of territory, including some of China's most fertile farmland, precious archeological sites, and highly polluted industrial sites. By the time it's all over, the Three Gorges Dam project will have caused between 1 and 2 million people to relocate to higher ground.

The Challenges

China's population is huge. By stressing (and sometimes unacceptably enforcing) one-child families, late marriages, and all kinds of birth control, China has reduced its population growth rate to a more manageable 0.6 percent. But even that figure, multiplied by the country's population of 1.3 billion, translates into an increase of nearly 8 million people per year.

Some of the worst famines in human history have occurred in China. Feeding such a huge population is still a major undertaking. China has more acres of farmlands, forests, and grasslands than almost any other country on Earth, yet its per capita acreage is about a third of the world's average—a worrisome statistic in a country whose major rivers are particularly prone to devastating floods and that is losing significant amounts of arable land to erosion and development. According to some estimates, China has lost as much as one fifth of its farmland since 1949.

Although China continues to modernize and privatize its economy at a remarkable pace, it has not been a smooth, continuous process. China's leadership has a tendency to backtrack periodically, reimposing centralized controls. In many ways, China suffers the worst of both its systems: the inefficiency and bureaucracy of the state controlled sector and the sudden windfalls and income disparities of unregulated capitalism. Unemployment and underemployment are huge, if somewhat concealed, problems in China, especially in rural areas. In 2006, the urban unemployment rate was 4.2 percent, but this does not include the millions who have already been sent home (at reduced wages) from closed or downsized state factories. The official figure also doesn't include members of the "floating" migrant population who have left the private rural sector in search of either work or a better life.

And finally, like Russia, China still needs to fully define its position in the world today. Modern China lives and dies by trade and it needs a cooperative global environment to stoke its massive economic engine. Retreating into hostile isolation is not an option any longer; neither is baiting its major trading partners. China has a host of international issues to resolve, including human rights, Taiwan (which it considers a "renegade" province), and how it handles Tibet and North Korea.

A Quick Look at China's People

About 93 percent of China's population is ethnic Han Chinese. China's 55 other minority groups make up the remaining 7 percent. The most numerous of these groups are the Zhuang, Hui, Uygur, Yi, Miao, Manchu, Tibetans, Mongolians, Buyi, and Koreans. Mandarin and Cantonese are the most commonly spoken of the dozen or so Chinese dialects. For 3,000 years, China has had only one written language, which not only eases communication but also contributes immeasurably to a common culture. Both internally and through the United Nations, China is pushing the use of Pinyin, a system of phonetic spelling, for personal and place names (Beijing for Peking and Mao Zedong for Mao Tse-tung, for example).

The communist government officially discourages religion and has periodically suppressed various beliefs and denominations. The most recent example was the campaign against Falun Gong, an extremely popular, and apolitical, religious movement that combines elements of traditional Chinese religions with martial arts. After the traditional Chinese religions, the most common religions in China are Christianity and Islam.

The Forgotten Republic of Mongolia

Mongolia occupies an isolated and desolate spot in central Asia. This lonely place is bounded on the north by Russia and on the east, south, and west by China. Though Mongolia is a large country (nineteenth largest in the world), it has a population of only 2.7 million people. With a density of only four people per square mile, Mongolia is the world's most sparsely populated country. Its capital is Ulan Bator, with a population of about 966,000.

The country's most productive region is its northern plateau. To the west are the Altai Mountains. In the south is Mongolia's most prominent feature, the Gobi Desert. The Gobi—the coldest, northernmost desert in the world—covers more than 500,000 square miles. This dry, rocky, foreboding place does not conform to the classic desert image: only 5 percent is covered by sand dunes.

Mongolia's climate is extreme—extremely cold in winter and extremely hot in summer. (Because it's in central Asia, no nearby large water body moderates climate extremes.) Most Mongolian people are rural nomadic herders. Its annual per capita GDP is only $1,770, which puts it in the same bracket as Gambia and just above Uganda.

As the Soviet Union was falling apart in 1990, pro-democracy voices were also making themselves heard in Mongolia. In 1996, the Mongols elected their first noncommunist government since 1921. But former communist party members maintained their grip on the nation's affairs and slowed the pace of reform. In the 2001 elections, the former communist party took control of both the presidency and parliament, although promising to implement much-needed reforms.

The Democratic People's Republic of North Korea

North Korea is home to about 23 million people, all of whom speak Korean. North Korea's climate is temperate and its terrain is mountainous. Its capital is Pyongyang, a city of more than 3 million people. North Korea is a Marxist, government-controlled state that makes China look liberal! At the end of World War II, the Korean peninsula was divided at the thirty-eighth parallel, with the northern part going to the Soviet Union and the southern part to the United States. In 1948, the northern half became the communist Democratic People's Republic (North Korea), and the southern half became the capitalist Republic of Korea (South Korea). Two years later, North Korea invaded South Korea and attempted a military unification. With help from the United States, North Korea was pushed back almost to the Chinese border; with Chinese help, South Korea was pushed back to the thirty-eighth parallel, where a no-man's land was set up in the 1953 truce.

The country's communist dictators have driven North Korea into poverty and starvation, while maintaining the fifth largest standing army in the world. After decades of economic mismanagement, North Korea's industrial base is a shambles and the country is heavily dependent on foreign food aid. A four-year famine in the late 1990s is thought to have claimed as many as 3 million lives. Although the food situation has eased in recent years, the North Korean population is still subject to malnutrition and deteriorating living standards.

In the News: The North Korean Nuclear Threat

Tension has existed between the two Koreas since the end of the Korean War, but has escalated dramatically since U.S. President George W. Bush, in his 2002 State of the Union message, lumped North Korea with Iraq and Iran in the "axis of evil." The North Korean dictator, Kim Jong-il, has spent most of his country's meager resources on his military and in developing nuclear weaponry and missiles. This is a particular concern, especially to Japan which is in close range, but also to the United States. Dictator Kim has boasted of his, so far unproven, ability to hit the United States with his missiles. This is an ongoing nuclear crisis, especially since North Korea joined the "nuclear club" by detonating a crude bomb in October 2006.

The Least You Need to Know

- With 1.3 billion people (one fifth the world's population), China is the world's most populous country and the third-largest country in area.

- China has the world's second largest economy.

- Unlike Russia and the former Soviet satellites in Eastern Europe, China is still a communist country politically.

- Staunchly communist North Korea lags significantly behind South Korea economically, and its nuclear weapons program is a source of global tension.

Part 4

A Regional Look: The Developing World

By definition, the developing countries are less wealthy and less technologically advanced than the developed countries (Part 2) and most of the transitioning countries (Part 3). Yet such distinctions can be deceiving. Many of the countries in this group are indeed desperately poor, verging on hopeless. But some are actually on the brink of joining the economic A-list, while others have enormous potential, thanks to their as yet untapped natural and human resources. It's also true that some very poor countries are extremely rich culturally or spiritually. And of course, some of the most beautiful places on Earth are to be found in the developing world.

"I hear they run around in circles to get exercise."

Chapter 18

Middle America: Bridging Two Continents

In This Chapter

- ◆ The Middle American bridge to the past
- ◆ The mainland connection
- ◆ Mexico: Middle America's heavy hitter
- ◆ A cruise through the enchanted Caribbean islands

The continent of North America includes two regions: developed Northern America (Canada and the United States) and developing Middle America (Mexico, Central America, and the islands of the Caribbean). Geographically, Middle America serves as a bridge between the rest of North America and South America, two giants that are both connected and separated by Middle America. Actually, Middle America is a dual bridge: the funnel-shaped mainland in the west (this is what physical geographers refer to as a land bridge) and the arc of islands in the east. The twin spans of the Middle American bridge are separated by the Gulf of Mexico and the Caribbean Sea.

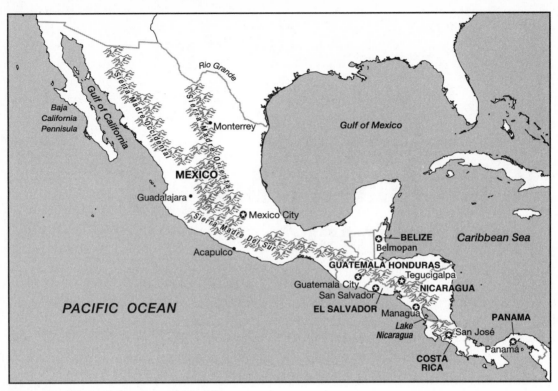

Mexico and Central America—physical features.

Empires of the Past

Mexico in Middle America and Peru in South America were the main hubs of Spain's vast New World empire, but they were also home to amazing cultures long before Europeans arrived.

The Amazing Maya

The Mayans stand out for the magnificence of their architecture and the sophistication of their science and art. The Mayan civilization developed in the lowlands of what are now Honduras, Guatemala, Belize, and the Yucatán Peninsula in Southeastern Mexico. At its peak, the Mayan population was between 2 and 3 million people. Today, descendants of the ancient Maya preserve their cultural and tribal traditions, with large numbers of them still speaking Mayan dialects rather than Spanish.

The Mighty Aztecs

About the time that the Mayan civilization had reached its peak, the civilization of the Toltecs was beginning to form around present-day Mexico City. The warlike Toltecs extended their domination south to the Yucatán Peninsula, where they conquered the declining Maya. The Toltec empire, however, was relatively short-lived; by 1200 C.E., it was in decay.

The Toltecs were supplanted as the overlords of central Mexico by the warlike Aztecs.

Terra-Trivia

Maize (corn) was of vital importance to the Maya and to all the peoples of the Middle American mainland. In addition to being their chief food crop, corn was believed to be the material out of which the gods created humankind.

The Spanish Empire in North America

After Hernan Cortez, the Spanish conquistador, conquered the capital of Tenochitlan in 1521, the Spanish imposed Catholicism and a harsh quasi-feudal economy on the defeated Native Americans of Middle America. They also extracted massive amounts of wealth from the silver mines of Mexico and from the forced labor of the indigenous population throughout the region. With the passage of time, a native-born creole elite began to feel greater allegiance to its American homeland than to the mother country of Spain, while the mixture of Spanish-born settlers and Indians resulted in the creation of a new mestizo, or mixed-race, population. By the beginning of the nineteenth century, resentment against Spanish rule was reaching a boiling point, and after a long struggle, Mexico and the Central American provinces declared their independence from Spain in 1821.

The Geography of Middle America's Mainland

The Middle American mainland bridge extends from the United States-Mexico border down to the narrow *isthmus* of Panama that connects to South America. Only 40 miles wide at its narrowest point, the isthmus gradually widens as it progresses in a northwesterly direction. The Middle American mainland has two major bulges on its eastern side: the eastern Nicaraguan and Honduran hump and the Yucatán Peninsula. The Baja California peninsula on the west juts south for more than 750 miles from the point where the Middle American bridge (now about 1,400 miles wide) joins with the rest of North America.

The mainland funnel has as many geographic contrasts as it does similarities. The great Rocky Mountain chain of Northern America continues south through Middle America under many local names (most variations on Sierra Madre) until it reaches the massive South American Andes Mountains. Most mountains in Middle America are volcanic in origin, and many are still active.

The combination of volcanic eruptions and earthquakes, which are also frequent in this region, make the Middle American mainland one of the more dangerous areas in the world. In most of the countries of this subregion, mountains run through the length of the country with highland plateaus in the middle and lowland coastal forested belts or plains to the west and east. The climate is affected by the altitude: the lowlands and coastal belts are tropically hot and extremely rainy (as much as 150 inches a year), while the highlands are temperate, much cooler, and less rainy.

The Middle American mainland funnel, more than 3,800 miles long, is anchored by giant Mexico in the north. At 762,000 square miles, it is by far the largest country in the region—almost twice the size of all the others combined—and the fourteenth largest in the world.

Mexico: Middle American Dynamo

Mexico and the United States have had difficulties. The Mexican-American War of 1846 to 1848 was triggered by the decision of Texas, which had broken away from Mexico in 1836, to join the United States. Mexico lost the war and was forced to cede what are now the states of California, Nevada, and Utah, most of Arizona and New Mexico, and parts of Wyoming and Colorado. The United States again intervened militarily in Mexico in 1916 during a period of revolutionary strife and civil war.

Mexico's twentieth-century history was dominated by a single political party, the PRI (from the Spanish for Institutional Revolutionary Party). In July 2000, 71 years of one-party rule came to an end with the election of President Vicente Fox of the opposition National Action Party. Six years earlier, Mexico had taken an equally giant step in the economic sphere by joining the United States and Canada in the North American Free Trade Agreement (NAFTA). This landmark and still controversial treaty will phase out all tariffs among the three nations over a 15-year period and help integrate Mexico's economy more fully with those of the two North American powerhouses.

Mexico, officially The United Mexican States, is composed of 31 states and a federal district. It has a population of more than 108 million people, and its largest cities are the enormous capital of Mexico City (with a metropolitan area population of more than 19 million, Mexico City is one of the three largest cities on Earth) and the industrial centers of Guadalajara and Monterrey. The majority of the population is mestizo (Amerindian-Spanish) or either pure or predominantly Amerindian. More than 90 percent of Mexicans are at least nominally Roman Catholic. The primary language of the country is Spanish, although many indigenous languages are also spoken. Mexico today is still plagued by high levels of poverty, underemployment, and inequitable income distribution. Yet it is also a major oil and gas exporter with a modernizing economy that has made extraordinary strides since the mid-1990s.

> ### Eye on the Environment
>
> Mexico City is one of the largest and most vibrant cities in the world, and a key center of Latin American economic and intellectual life. But it is also a city plagued by severe environmental issues—a massive population, terrible air and water pollution, the gradual sinking of heavy buildings due to the lowering of the water table, and an extremely high risk of earthquakes.

The Nations of Central America

South of Mexico on the Middle American land bridge is the subregion of Central America, consisting of seven countries: Guatemala, Belize, Honduras, El Salvador, Nicaragua, Costa Rica, and Panama.

Spanish is either the official or the main language of all of the Central American countries, except Belize, where it's English. The countries' populations are generally of mixed race, either mestizo (European and Amerindian) or mulatto (European and black African). Most Central Americans are Roman Catholic.

Central America declared its independence from Spain on September 15, 1821, only to be annexed by the short-lived Mexican Empire. The Central American provinces opted to separate from Mexico and form the United States of Central America. The union began to dissolve in 1838.

Guatemala

With more than 13 million people, Guatemala is the most populous country in Central America. With more than 5 million people, the capital, Guatemala City, is Guatemala's and Central America's largest city. A poor, mostly agricultural country, Guatemala has had a tormented history marked by civil strife and military dictatorships. A peace agreement signed in 1996 ended a 36-year guerilla war—the longest in Latin American history—that left some 100,000 people dead (200,000 according to some estimates) and created about a million refugees.

Belize

Belize is the only Central American country with a British background (it was called British Honduras until 1981 when it gained its independence from Britain) and the only one with no Pacific coastline. The second-smallest country in the subregion, it's also one of the least populated, with just over 290,000 people. The economy is heavily dependent on tourism, followed by agriculture (sugar cane, citrus fruits, and bananas).

Honduras

Honduras, the second-largest country in Central America with more than 7 million people, is at the bottom in terms of wealth along with Nicaragua. Honduras has had a history of dictators, revolutions, and coups. During the 1980s, the country served as a base for anti-Sandinista guerrillas fighting Nicaragua's Marxist government and assisted the Salvadoran government in its fight against leftist guerrillas. The capital Tegucigalpa is a city of more than 1 million people. The economy of Honduras is based on agriculture; coffee and bananas are the main exports.

El Salvador

El Salvador, the smallest country in Central America in land area, has a population of more than 7 million and is the only one with no Caribbean coastline. Following independence in 1821, El Salvador was involved in several wars with its neighboring countries. The first few decades of this century saw some stability and growth in the economy, only to be followed by a series of strong but oppressive dictatorships. A costly 12-year civil war finally came to an end in 1992. Honduras, El Salvador, and Nicaragua were all badly affected by Hurricane Mitch in 1998, and powerful earthquakes and drought struck El Salvador in 2001. The capital San Salvador has a population of more than 2 million residents. A mainstay of the economy is money sent home by Salvadorans working in the United States. The U.S. dollar is the country's legal tender.

Nicaragua

With a per capita GDP of only $2,900, Nicaragua is not only the largest country in Central America but also the poorest. Corruption and political upheaval are largely to blame for the country's economic difficulties. The Marxist Sandinista regime came to power in 1979, but was opposed throughout much of the 1980s by the U.S.-sponsored guerrilla fighters known as the *contras* ("those against" in Spanish). Beginning with the free elections of 1990, the Sandinistas were repeatedly defeated at the polls, until the November 2006 re-election of the Sandinista leader, Daniel Ortega, as president. President Ortega returned to power promising a gentler "socialism," but politics aside, he probably won simply because nothing changed for the poor under the contras, so they voted to try Ortega's new plan. The capital of Managua with 1.4 million people sits aside Lake Managua in the lowlands that connect to Central America's largest lake—Lake Nicaragua.

Costa Rica

Costa Rica's per capita GDP of $11,400 is the highest in Central America. Its indigenous people resisted colonization, and, with no obvious wealth, Spanish subjugation came late and then only indirectly through Nicaragua. Costa Rica achieved independence in 1821, and its transition to democratic rule was relatively smooth. With no army and a thriving middle class, the entire country enjoys a standard of living that's quite high by Middle American standards.

Panama

In many ways, Panama is a creation of the United States. With U.S. backing, it seceded from its South American neighbor, Colombia, and promptly signed a treaty with the United States allowing it to build the Panama Canal and take possession of the Canal Zone. Panama's history has been plagued by dictators and coups. U.S. forces invaded Panama in 1989 to oust the most recent dictator, General Manuel Noriega, who was subsequently tried and convicted on drug and money-laundering charges. Panama's stable, service-based economy is the second-strongest in Central

America after Costa Rica. South of the canal is the "Darien Gap"—an area of some of the world's densest rainforest that interrupts the Pan American Highway that would otherwise link the southern tip of South America to Fairbanks, Alaska.

> ### Geographically Speaking
>
> Generally recognized as one of the greatest engineering feats of all time, the Panama Canal was built between 1905 and 1914 by the U.S. Army Corps of Engineers. The canal is 40 miles long and has lock chambers that lift and lower ships 85 feet during the passage between oceans. By going through the canal instead of around South America, a ship sailing from the east coast of the United States to Japan saves about 3,000 miles. Under the terms of a 1977 treaty, the United States ceded full control of the canal to Panama on December 31, 1999.

Middle America's Caribbean Island Paradises

Most of these islands have a mulatto (mixed African/White) population, mestizo (White/Amerindian), or African population:

- **West Indies:** The catch-all term for all the island groups in the area: the Bahamas, Turks and Caicos Islands, Greater Antilles, Cayman Islands, Lesser Antilles, and Netherlands Antilles

- **Greater Antilles:** The big islands of Cuba, Hispaniola (Haiti and the Dominican Republic), Jamaica, and Puerto Rico

- **Lesser Antilles:** The arc of small islands in the Eastern and Southern Caribbean: the Leeward Islands, Windward Islands, and Netherlands Antilles

- **Leeward Islands:** The northern half of the island arc that is sheltered from the trade winds: the Virgin Islands, Saint Eustatius, Montserrat, Saba, Antigua-Barbuda, Saint Kitts-Nevis, and Guadeloupe

- **Windward Islands:** The southern half of the island arc: Dominica, Martinique, Saint Lucia, Barbados, Saint Vincent and the Grenadines, Grenada, and Trinidad and Tobago

- **The ABC Islands:** Just off the northern coast of South America, the Dutch islands of Aruba, Bonaire, and Curaçao

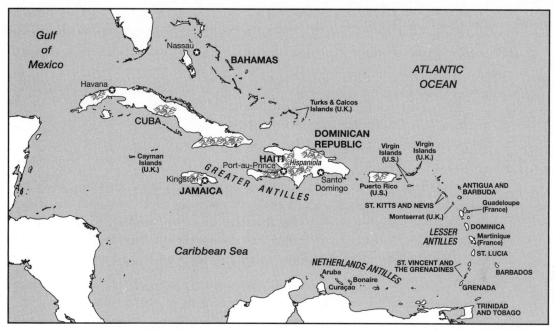

Islands of the Caribbean—physical features.

The Greater Antilles: The Big Islands

This subregion of Middle America consists of four main islands: Cuba, Hispaniola (with the countries Haiti and the Dominican Republic), Puerto Rico, and Jamaica. Cuba is by far the largest island in the Greater Antilles group. On the east, Cuba is separated from Hispaniola by the Windward Passage. Hispaniola, the second-largest island, contains two nations: Haiti, which occupies the western third of the island, and the Dominican Republic, which occupies all the rest. South of Cuba is the third-largest island, Jamaica. The smallest island in the group, Puerto Rico, lies east of Hispaniola.

These four islands have in common their large size (compared to the Lesser Antilles) and they're all close together in the north Caribbean. All the islands of the Greater Antilles were originally colonized by Spain. However, Britain conquered Jamaica in 1655, and in 1697, Spain officially ceded the western third of Hispaniola, today's Haiti, to France. What is now the Dominican Republic remained in Spanish hands until 1795, when it, too, was ceded to France. The Dominican Republic is still Spanish speaking, as are Cuba and Puerto Rico, which remained Spanish colonies until 1898.

Although similar geographically in many ways, the Greater Antilles are also full of contrasts. Cuba, for example, is relatively flat and only about one-quarter mountainous, while the other islands are very mountainous. All have tropical or semitropical climates, but there are climatic contrasts as well. In Jamaica's rain forest, the amount of rainfall can exceed 200 inches per year, while parts of Haiti get only 20 inches of rain annually. All the islands in this group are blessed with miles of spectacular beaches that have made them meccas for tourists from around the world.

Cuba

Cuba has a population of more than 11 million people, all Spanish-speaking and most with a mixed Spanish and black African or Amerindian heritage. Its capital and largest city is Havana, once the Paris of the Caribbean, now in a serious state of disrepair. Before the victory of Fidel Castro's communist revolution in 1959, Cubans were almost all Roman Catholic, but now only 40 percent are, and half declare themselves nonreligious.

Cuba's economy has been struggling to regain its footing after the collapse of the Soviet Union and the loss of Soviet aid (about $5 billion a year) in 1991. Sugar, nickel, and tobacco (those famous Cuban cigars) are the country's main exports. Although the U.S. embargo against Cuba, which Castro blames for most of his nation's ills, prevents most American citizens from traveling to Cuba, the island is a popular destination for European tourists, and the tourism sector had been growing steadily until the post-September 11, 2001, slump.

Geographically Speaking

A particular bone of contention between the Castro government and the United States is the U.S. naval base at Guantánamo Bay in the southeastern tip of Cuba. The United States established the base at the end of the Spanish-American War in 1898 and leased it permanently in 1903, in a deal that Castro considers to have been illegitimate.

Guantánamo has been used at various times to house Cuban and Haitian refugees. In 2002, the United States began housing suspected Taliban and al-Qaeda prisoners taken in Afghanistan in a Guantánamo base prison called Camp X-ray. Because the base is in Cuban territory, U.S. prisoners held there do not enjoy the same constitutional rights they would if they were being held in U.S. territory.

After more than 40 years in power, Castro still dominates the Cuban story. A critical question for Cubans, on the island and in exile in the United States, is what happens after Castro?

Haiti

In the eighteenth century Haiti was the source of incalculable wealth for its French overlords. But the Haitian sugar economy was built on the backs of hideously oppressed and ill-treated slaves. In the only successful slave uprising in history, Haitians overthrew the plantation owners and went on to defeat the armies of Napoleon to become the world's first independent black republic. Haiti's post-independence story is far less glorious. Wars, uprisings, and cruel dictatorships have turned Haiti into one of the poorest countries in the world.

Haiti has about 8.5 million people, about 95 percent of whom are black. The capital and main city is Port-au-Prince with more than 1.2 million inhabitants. French and Creole are both official languages. Although most Haitians are Roman Catholic, many practice voodoo, a blend of Catholicism and African religions.

Nearly 80 percent of Haiti's population lives in extreme poverty, barely eking out a living from subsistence farming. To make matters worse, foreign aid was essentially cut off in 2000 after a legislative election that was marred by irregularities. Of the many crises facing Haiti, poverty, disease, and ecological destruction top the list. Infant mortality rates are high, life expectancy is short (under 50 years), and the country has the highest HIV and tuberculosis infection rates in the Western Hemisphere. On the environmental front, 99 percent of Haiti is now deforested and tropical rains have washed away much of the soil. The result is a land that's denuded—in many areas, down to bedrock level. What little soil remains is overworked and is used for growing export crops rather than life-sustenance crops.

Dominican Republic

The Dominican Republic shares the island of Hispaniola with Haiti, which ruled it from 1822 to 1844. The period of Haitian rule left a bitter legacy of mutual antagonisms and prejudices that persist to this day.

The Dominican Republic has also suffered from political violence, foreign (U.S.) intervention, and cruel dictatorships. The Dominican Republic has a population of more than 9 million people who speak Spanish, have a mixed black and Spanish heritage, and are predominantly Roman Catholic. Its capital and largest city is Santo Domingo with more than 2 million residents, the oldest European settlement in the New World. Although the Dominican Republic is known primarily as an exporter of sugar, coffee, and tobacco, tourism has overtaken agriculture as its main income earner.

Jamaica

After more than 100 years of Spanish rule followed by more than 300 years of English rule, Jamaica finally attained its independence in 1962. Since its independence, the island has experienced periods of political unrest and crime waves that have threatened the vital tourism industry.

Jamaica has 2.7 million inhabitants, consisting mostly of blacks (91 percent) and people of mixed heritage (7.3 percent), plus small English, South Asian, and Chinese minorities. The Jamaican capital is Kingston.

Most Jamaicans speak the official language, English, or a local English dialect that incorporates African, Spanish, and French elements. The majority of Jamaicans are Protestants, although the island is also the birthplace of Rastafarianism, a religious sect whose members believe in a prophetic return of blacks to Africa.

The two pillars of the Jamaican economy are tourism and bauxite. After a long recession in the second half of the 1990s, Jamaica has been experiencing negligible growth in the first years of this century as tourism continues to be affected by the aftermath of 9/11 and a slow-to-recover global economy.

Puerto Rico

Puerto Rico is by far the wealthiest nation in the Greater Antilles, and the only one that is not independent. Its capital and largest city is San Juan. Its population of nearly 4 million people is Spanish speaking and mostly Roman Catholic. Puerto Ricans are of predominantly white Spanish descent; about 8 percent are black and 11 percent are of mixed heritage. Industry, especially pharmaceuticals, and tourism have overtaken agriculture as the prime movers of Puerto Rico's economy.

Puerto Rico was a Spanish colony until the Spanish-American War in 1898, when it was ceded to the United States, under whose control it has remained. Puerto Ricans were granted U.S. citizenship in 1917. They have been electing their own governors since 1948 and handling their internal affairs since 1952. As a commonwealth of the United States, Puerto Rico does not have voting representation in the United States Congress nor can island residents vote in U.S. presidential elections. (Puerto Ricans who move to the mainland can vote for president.) On the other hand, Puerto Ricans do not pay federal taxes. The future status of Puerto Rico—statehood, independence, or commonwealth—has been a matter of much debate for decades. In 1967, 1993, and again in 1998, the people of Puerto Rico voted to maintain their commonwealth status.

The Lesser Antilles: The Small Island Arc

The fourth subregion of Middle America is an archipelago (a large group of islands) that extends in a grand arc all the way down to South America. Consisting of the Lesser Antilles and the Netherlands Antilles, the subregion includes nine independent countries, many possessions and territories, and almost countless islands, islets, and cays (tiny islands of rock, sand, or coral).

The Lesser Antilles are all small in both population and size. Most of the islands are lush and some have rain forests. Almost all of them have golden white (or in some cases, black) sand beaches and coral reefs. The Lesser Antilles are actually the tops of underwater mountains; some have active or recently active volcanoes.

In most of the islands of the Lesser Antilles, the population is predominantly black (descendents of Africans brought over as slaves to replace the decimated Amerindians) or of mixed European and African ancestry. An exception is Trinidad and Tobago, where immigrants from Northern India and their descendents make up more than 40 percent of the population. Trinidad and Tobago is also unique in that it has a major oil and natural gas industry and tourism is less important than in most of the other Caribbean islands. Petroleum refining and transshipment are also key components of the Netherlands Antilles economy. Like the Greater Antilles, the islands of the Lesser Antilles are a tourist paradise for Americans, Europeans, and everyone else seeking tropical island warmth and beauty.

They are often sorted out by who their colonial masters used to be:

♦ **British:** The British Virgins, Antigua-Barbuda, Saint Kitts-Nevis, Montserrat, Barbados, Saint Lucia, Saint Vincent and the Grenadines, Grenada, and Trinidad and Tobago. Although the Virgins and Montserrat are still dependencies of the United Kingdom, the others are independent countries that elect their own heads of government but recognize Queen Elizabeth II as their sovereign.

♦ **Danish:** The U.S. Virgin Islands. In the seventeenth century, the Virgin Islands were divided between Britain and Denmark. The Danes sold their part to the United States in 1917.

♦ **Dutch:** Part of the Kingdom of the Netherlands, the Netherlands Antilles used to consist of the ABC islands—Aruba, Bonaire, and Curaçao—and three of the Leeward Islands (actually 2½ islands): Saba, Saint Eustatius, and Sint Maarten (half of an island shared with France). In 1986, Aruba pulled out of the Netherlands Antilles to become a separate part of the Dutch kingdom.

◆ **French:** Martinique and Guadeloupe are full overseas departments of France, not colonies or dependencies. Guadeloupe is actually an archipelago of nine inhabited islands, including Saint Martin, the other half of the island shared with the Dutch. Dominica has been a true political football: it started out Spanish, then became French, then English, and finally independent.

The Least You Need to Know

◆ The Middle American region has four primary subregions: Mexico, Central America, the Greater Antilles, and the Lesser Antilles.

◆ The Middle American mainland was home to the great empires of the Maya and the Aztecs.

◆ Mexico is the largest and most politically and economically important country in the region.

◆ The Panama Canal connects the world's two largest oceans via the thin isthmus of Panama in Central America.

◆ The Greater Antilles include the four largest Caribbean islands: Cuba, Hispaniola, Puerto Rico, and Jamaica.

◆ The Lesser Antilles consist of many small islands, where tourism has mostly supplanted sugar production as the main industry.

Chapter 19

South America: The Iberian Legacy

In This Chapter

- ◆ Surveying South America's troubled history
- ◆ Exploring the continent's countries and cities
- ◆ Traveling down the Amazon and up the Andes
- ◆ Meeting South America's mix of peoples
- ◆ A continent at a crossroads

South America is the fourth-largest continent, with 12 percent of the earth's landmass. It is much farther east than most people realize. Longitudinally, virtually the entire continent lies east of Florida. Its eastern bulge is less than 1,700 miles from the western bulge of Africa, whereas the Atlantic Ocean measures more than 3,500 miles from New York to Lisbon.

The distance from Western South America to Australia is almost twice as much as from California to Japan. The equator passes through the northern part of South America. From East Coast to West Coast, the continent measures 3,000 miles; from north to south, it's 4,600 miles. South America is 900 miles south of Florida but only 600 miles north of Antarctica.

South America.

Peopling the Continent

The most commonly held theory about when people arrived in South America is that East Asian peoples crossed the Bering land bridge (when it connected Asia with Alaska) and then fanned out east and south, eventually passing down through the

Middle American land funnel into South America. The archeological evidence indicates that humans had reached the southern tip of South America by 9,000 to 8,500 B.C.E. The discovery of coastal settlements in Peru dating back 12,000 years opens up the possibility that South America was first inhabited by people traveling by foot and, in stretches, by boat along the west coast of the Americas. Before the Europeans "discovered" the Americas, hundreds of native tribes lived, hunted, herded, and farmed in South America.

Terra-Trivia

Inca architects built dry stone walls (which still stand) to terrace hillsides. Archaeologists still look with amazement at Inca masonry: despite the use of only massive stone blocks and no mortar, the joints are so fine that an index card cannot fit between the stones.

The Inca Empire

The Incas first arrived in the Andean highlands of Peru in about 1150 C.E., and over the next 300 years established their domination over the peoples living in the Valley of Cuzco. The Inca Empire included all of present-day Ecuador, most of Peru and Chile, and parts of Bolivia and Argentina. Its territories measured 2,500 miles long by 500 miles wide and contained almost 20 million people. This vast empire was later conquered by Spanish Conquistadors in the early 1500s.

Exploring and Exploiting the New Continent

Spain exploited the continent and decimated the native population. Soon, other European powers began to claim lands and establish settlements in the "Spanish territories of the new world."

The Spanish and Portuguese set up huge agricultural settlements and imported millions of black Africans to labor on their plantations.

This slave system of exploitation was matched by the European states' exploitation

Terra-Trivia

In the seventeenth century, the Dutch and British both occupied what is now Suriname. In 1667, the British ceded their share to the Netherlands in return for the Dutch colony of New Amsterdam, today's New York City—nice trade!

of their colonies for almost 300 years. In the early 1800s the colonies began to rebel against European domination, and slave revolts and calls for freedom by the black slaves quickly followed colonial independence.

The South American Countries and Their Major Cities

Following is a brief overview of South America's 12 nations (and one French department).

Brazil

Brazil is South America's largest (and the world's fifth largest) country, in terms of both population and area. With about 188 million inhabitants, Brazil is home to more than half of all South Americans. The country has 13 cities with a population of more than 1 million. The two largest are São Paulo (South America's most populous city) with 11 million people (more than 19 million in its metropolitan area), and Rio de Janeiro, with almost 6 million (more than 11 million in its metropolitan area). Brazil's capital, Brasília, is relatively small in comparison with a population of more than 2.3 million people. The country's twentieth-century history was marked by repeated military interventions and government corruption. Brazil today has South America's biggest economy, but grossly unequal income distribution remains a problem. In 2003, Brazilians elected their first president from a working-class background.

Colombia

With more than 45 million inhabitants, Colombia is South America's second most-populous country. Unique among South American countries, Colombia has coastlines on both the Pacific Ocean and the Caribbean Sea. More than 7 million people live in the Colombian capital of Bogotá. Its second largest city, Cali, has more than 2 million inhabitants; Medellín and Barranquilla each have populations of more than 1 million people. A focal point of the international drug trade, Colombia has also suffered for decades from an increasingly violent anti-government insurgency.

Terra-Trivia

The dream of South America's great liberator, Simón Bolívar, was to unite the entire continent as one nation, but that was never to be. Even the Republic of Gran Colombia, which he founded in 1819, lasted barely a decade. By 1830, the republic had broken up into the three modern-day nations of Colombia, Venezuela, and Ecuador.

Argentina

Argentina is South America's third most populous country (with more than 40 million inhabitants) and its second-largest in land area. Its capital, Buenos Aires, birthplace of the tango, has nearly 3 million people, but its greater metropolitan area is home to nearly a third of the country's total population. Cordoba and Rosario are the other Argentine cities with more than a million inhabitants. Argentina's vast natural and human resources make it a natural candidate for the economic leadership of South America, but a post-World War II history of authoritarian rule, military dictatorship, and political corruption have kept this rich nation from fulfilling its potential. Although Argentina has consolidated its democratic institutions following the military dictatorship and "dirty war" of the late 1970s and early 1980s, its economy struggled through the 1990s and nearly collapsed in 2000 and 2001.

Since mid-2002, Argentina's economy seems to be on the mend, but remains fragile. Despite recent setbacks, Argentina still has South America's highest per capita gross domestic product (GDP).

Peru

Peru is home to about 28 million people, just over half of whom are Incas and other Amerindians living in the Andean highlands. The remaining population consists mostly of mestizos, or mixed Amerindians and whites (32 percent), whites (12 percent), and blacks and Asians (about 3 percent) living on the coastal plain. Peru's capital and largest city is Lima, with 6.8 million residents in the greater metropolitan area. Like its neighbors, post-World War II Peru has experienced its share of military dictatorships, anti-government insurgencies, and economic problems. In the 1990s, Peruvian president Alberto Fujimori achieved notable success in suppressing the violent Marxist rebel group known as Shining Path, but his rule grew increasingly authoritarian. In 2000, Fujimori was ousted and forced into exile in Japan. Political instability had a negative impact on Peru's economy, although as of 2002 the country is experiencing strong growth. Mining and the extraction of petroleum and gas are important sectors of the Peruvian economy. Peru is also the world's largest grower of coca, from which cocaine is refined.

Venezuela

Venezuela is a country of nearly 27 million people, 3.5 million of whom live in the modern capital of Caracas. The cities of Maracaibo and Valencia both have populations of more than a million people. Venezuela has the largest oil reserves in the

Western Hemisphere and the sixth largest in the world. Its economy, however, is overly dependent on the petroleum sector and therefore highly vulnerable to oil-price fluctuations. In recent years, Venezuela's economy and society have been rocked by political instability under the divisive regime of leftist president Hugo Chávez.

Chile

Chile is 2,650 miles long from north to south, but averages only 110 miles wide from east to west. Its population of 16.5 million people is almost entirely mestizo (mixed white and Amerindian) or white. Full Amerindians make up about 10 percent of the population. Chile's capital, Santiago, has more than 5.5 million people in its metropolitan area, or about one third of Chile's total population. Chile's history over the last 30 years has been particularly traumatic. The overthrow and violent death of leftist president Salvador Allende in 1973 ushered in a repressive 17-year military dictatorship under General Augusto Pinochet. Economic reforms begun under Pinochet and continued under the democratic governments that succeeded him have established Chile as one of South America's strongest and most stable economies. In December 2002, Chile signed a NAFTA-like free-trade accord with the United States.

Ecuador

Ecuador straddles the equator. The country's population of 13.4 million is mostly mestizo or pure Amerindian. The capital, Quito, located high in the Andes, has 1.8 million people in its metropolitan area. Coastal Guayaquil is even larger, with 2.6 million people in its metropolitan area. The famous Galapagos Islands belong to Ecuador. Ecuador was once the northern part of the Inca empire, but its modern relationship with its southern neighbor, Peru, has been far from smooth. During the twentieth century, the two countries fought several wars over territory, with Ecuador usually on the losing side. A treaty signed in 1999 finally ended the longstanding border dispute. The pillars of Ecuador's unstable economy (the U.S. dollar became the nation's currency in 1999) are petroleum and agriculture. Despite recent reforms, the Ecuadorian economy remains weighed down by budget deficits and a huge foreign debt.

Bolivia

Landlocked Bolivia (it lost its only corridor to the sea in a war with Chile) has a population of 8.4 million people, most of whom are either Amerindian (Quechua or Aymara) or mestizo (mixed white and Amerindian). It has two capitals: Sucre, the legal capital, and La Paz, the administrative capital and seat of government. La Paz has a population of around 1 million people and Sucre is one fourth that size (Bolivia's largest city is Santa Cruz de la Sierra with around 1.5 million people). Bolivia's history is a long, sad tale of coups, political unrest, and poverty. Although Bolivia has major natural resources and has made moves to reform and privatize its economy, it remains South America's poorest and historically most unstable country.

Paraguay

South America's other landlocked country, Paraguay has a population of more than 6 million, with more than 1.6 million living in and around the capital of Asunción. Almost all of Paraguay's people are mestizo. Paraguay has consolidated its democratic institutions since the overthrow of the cruel 35-year military dictatorship of Alfredo Stroessner, but its economy is held back by corruption, political uncertainty, and debt. Nearly half of Paraguay's work force is engaged in agriculture, often on a subsistence level.

Uruguay

Uruguay has a population of more than 3 million, most of whom are white and half of whom live in or around the capital of Montevideo. Like so many other South American nations, Uruguay was ruled by military dictators from 1973 to 1985, when democratic government was restored. Like neighboring Argentina, Uruguay has a strong agricultural sector and a well-educated workforce, neither of which prevented its economy from going into near free fall between 1999 and 2002, mostly as a result of Argentina's economic woes. The Uruguayan situation stabilized somewhat in 2003, but full recovery still seems distant.

The Guianas

The Guianas were all plantation colonies, more closely resembling the Caribbean colonies than the South American ones. The largest of the three is Guyana (the former British Guiana), with Dutch and English roots and a population of about 750,000 people, most of whom are East Indians (whose ancestors came from Northern India) or blacks. Guyana's capital is Georgetown. Its economy is based primarily on agriculture and bauxite mining.

The next-largest of the Guianas is Suriname (the former Dutch Guiana), with Dutch and English roots and about 450,000 people, most of whom are of South Asian ancestry, creole (mixed white and black), Javanese, and black. Suriname's capital is Paramaribo. Since gaining its independence from the Netherlands, Suriname has alternated between military and civilian governments. Its economy is heavily dependent on bauxite mining.

The smallest of the Guianas (and South America's smallest "state") is French Guiana, with only about 186,000 people, most of whom are either black or mixed black and white (whites make up about 12 percent of the population; East Indians, Chinese, and Amerindians, another 12 percent). French Guiana is an overseas department of France; its capital is Cayenne. French Guiana's economy is based on fishing, forestry, and subsidies from France.

Physical Facts

A glimpse at the following map of South America's physical features shows that the continent is dominated by two geographic wonders: the mighty Amazon River and the Andes Mountains. With its main trunk just south of the equator flowing from west to east, the Amazon dominates the northeastern portion of the continent. Running along the entire western edge of South America is the spectacular Andes Mountain chain, which acts as a huge wall dividing the continent's low interior areas from the vast Pacific Ocean.

The World's Largest River

Although the Amazon is probably second to the Nile in length (the Amazon is about 4,100 miles long while the Nile is 4,145 long), it's by far the world's largest river. The drainage basin of the Amazon encompasses more than 2.7 million square miles, or about 40 percent of South America's total surface area. Its discharge into the Atlantic Ocean measures more than 50 billion gallons per second. In one day, that's enough to meet the needs of all U.S. households for five months.

The Amazon changes course every year. With its more than 2,000 tributaries, a massive drainage system, and an abundant rainy season, the river rises annually an average of 65 feet and overflows the jungle on each side as much as 30 miles. This phenomenon makes the Amazon almost 60 miles wide at spots in the rainy season. When the river subsides during the dry season, its location is altered in many areas.

South America—physical features.

The High Andes

The Andes chain is one of the world's greatest mountain systems. It is 4,500 miles long and ranges from 50 to more than 400 miles wide. It reaches heights of as much as 22,831 feet (Mount Aconcagua, in Argentina, is the highest mountain in the Western Hemisphere). The snow line of the Andes ranges from 4,000 feet in the south to 17,000 feet at the equator. The tree line ranges from a southerly 3,000 feet to an equatorial 12,000 feet.

The Andes contain many lofty peaks, including 50 higher than 20,000 feet. Located at the juncture of the Nazca and South American plates, the Andes Mountains are subject to frequent, and often violent, earthquakes and volcanic eruptions. Of the 13 South American countries, only Paraguay, Uruguay, and the Guianas are considered non-Andean.

South America's Environments

Tropical rain forests dominate the Amazon basin and are the largest of their type in the world.

In Northern South America is an area of grassland and open dry forests called the llanos of Colombia and Venezuela. In the continent's southern cone are the famous pampas, home of the Argentine cowboys known as gauchos. This landscape of tall grass extends from central Argentina into Uruguay. In addition to being prime cattle country, the pampas are also a major grain-producing region.

South of the pampas is another major Argentine livestock area called Patagonia. With its higher elevations, cooler temperatures, and drier climate, Patagonia is ideally suited to sheep ranching. And finally, in a narrow strip west of the Andes in Northern Chile is the Atacama Desert, one of the driest places on Earth. Because the desert is on a 2,000-foot-high plateau and is favored by cool winds blowing off the Humboldt Current, its average year-round temperature is only 65 degrees F.

The Waters

With a surface area of 3,220 square miles, Lake Titicaca, between Bolivia and Peru, is South America's largest lake. (And at 13,000 feet above sea level, it is also the highest navigable lake in the world.) At 5,000 square miles, Venezuela's Lake Maracaibo is larger than Titicaca, but it's not a true lake. It's actually a bay connected to the Gulf of Venezuela by a narrow channel. Other large lakes dot the continent, many of them a result of human-made river dams.

In addition to the Amazon, South America has several long and locally important rivers. The Orinoco River in the north, more than 1,700 miles long, ends in a vast delta on the Atlantic Ocean. In the south, the 2,450-mile Paraná River flows from south-central Brazil. It has a confluence with the Paraguay River and then unites with the Uruguay River to form the Río de la Plata *estuary* on the Atlantic Ocean.

A few more physical features are important. In one of Venezuela's remotest spots, almost impossible to get to by land, are the Angel Falls. At 3,212 feet, they're the world's highest falls (compare them with Africa's Victoria Falls at 400 feet and Niagara Falls on the United States-Canada border, at 167 feet). On the Brazil-Argentina border are

def•i•ni•tion

An **estuary** is an inland arm of the sea that meets the mouth of a river. It also refers to the area where the river's current meets the sea's tides.

the Iguaçu Falls, one of South America's great natural wonders. In the dry season, the river drops in two 200-foot-high crescent-shaped falls. In the wet season, however, it becomes one gigantic fall about two miles wide (compare this with Victoria's one-mile width and Niagara's combined width of two thirds mile).

To South America's extreme south lies the Strait of Magellan. The strait connects the Atlantic and Pacific oceans by winding through South America just north of the southerly archipelago, Tierra del Fuego. It measures 331 miles long and only 2 to 15 miles wide. Although the strait is difficult to navigate, it enables ships to avoid the hazardous winds, rains, currents, and waves off Cape Horn at the southern tip of South America.

A Look at South America's Weather

South America's northern area is all tropical; Southern Brazil, half of Paraguay, almost all of Argentina, much of Chile, and all of Uruguay lie south of the Tropic of Capricorn (about 23½ degrees south latitude) and are temperate. For most of South America's tropical zone, the yearly temperature is 75 to 85 degrees F; it's cooler in the Brazilian Highlands, however, and even cooler in the Andes. In the temperate zone, winters range from the 60s to the 30s as you head south. Again, it's always cooler in the Andes. Down near Cape Horn, the average temperature is less than 32 degrees F year-round.

South America's People

South America has 12 percent of the earth's landmass but only 6 percent of its people. The large majority of these people live within 500 miles from a coast, particularly the east, central east, and northwest coasts. The vast interior is sparsely inhabited. Pure Amerindians live primarily in former Inca lands—Ecuador, Peru, and Bolivia. Mestizos (people of mixed Amerindian and European heritage) are found in the Inca lands, surrounding Amerindian populations, as well as in Chile, Paraguay, Colombia, Venezuela, and in smaller numbers throughout the continent. People of South Asian ancestry are the largest ethnic group in both Guyana and Suriname.

African blacks, brought to South America as slaves, live primarily in the Guianas, Brazil, and Venezuela, with smaller populations in Peru, Ecuador, and Colombia. Mulatto peoples (those of black African and European stock) live mainly in Venezuela, Brazil, the Guianas, and Colombia. Whites dominate only in Argentina and Uruguay, where they represent more than 85 percent of the total population. They also make up a significant part of the populations of Brazil, Chile, and Venezuela, and less than 20 percent of the populations of Peru and Colombia.

Portuguese is spoken in Brazil, English in Guyana, Dutch in Suriname, French in French Guiana, and Spanish everywhere else. Not surprisingly, many indigenous languages are also spoken. In Paraguay, 90 percent of the population speaks Guaraní. In Bolivia, Quechua and Aymara are both official languages along with Spanish, which about 40 percent of the people don't speak. Quechua is also an official language of Peru and is common in Ecuador. In Guyana, you hear Hindi and Urdu, and in Suriname, a creole language called taki-taki.

Roman Catholicism is dominant everywhere in South America, except in Guyana, where it's third to Hindu and Anglicanism. Because of South America's great diversity of ethnic groups and languages, Roman Catholicism is the continent's most universal unifying cultural force.

The Least You Need to Know

- ◆ South America is home to the world's largest river, the Amazon, and one of the world's greatest mountain chains, the Andes.

- ◆ South America's Andean region gave rise to the great Inca Empire.

- ◆ Brazil, with Portuguese, black African, and Amerindian roots, accounts for half of South America's land mass and population. The rest of the region has mainly Spanish and Amerindian roots.

- ◆ South America has had a troubled political and economic past, and despite recent stabilization and progress, it isn't out of the woods yet.

20

Saharan Africa: The Sands of Time

In This Chapter

- ◆ Navigating the Nile
- ◆ Discovering the Maghreb
- ◆ Traversing the transition zone of the Sahel
- ◆ Heading out on the Horn of Africa

The vast region of Saharan Africa encompasses the whole of Northern Africa. It is bounded on the west by the Atlantic Ocean; on the north by the Mediterranean Sea; on the east by the Suez Canal, the Red Sea, and the Indian Ocean; and on the south by the northern tier of sub-Saharan African countries, from Senegal to Kenya.

Although the region is made up of 14 countries, it is named for and dominated by the gigantic Sahara Desert, by far the world's largest. Saharan Africa's other major geographic features include the Atlas Mountains in the northwest; the great Nile River, Egypt's lifeline, flowing through Egypt and Sudan; and the Niger River in the southwest.

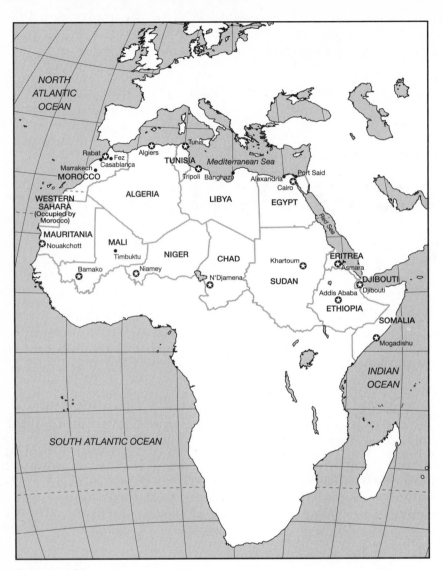

Saharan Africa.

Saharan Africa is united by the Islamic faith. The majority of the region's people, whether Arab, Berber, or black African, are Muslim. The northern tier nations bordering the Mediterranean are 95 to 99 percent Muslim. In the Sahel, indigenous religions and Christianity are practiced by minorities of varying sizes.

The Sahara is even larger than the region named for it. The desert extends beyond the Red Sea, across Arabia and the Persian Gulf, through Iraq, and into Iran. It measures more than 3.5 million square miles and includes about 80,000 square miles of oases, or fertile patches. Its daytime temperature in the shade can reach 130 degrees F, with nights falling into the 30s. Average yearly rainfall is less than five inches, and some areas get no rain at all for several years.

The Shifting Sands of Humankind

This region was home to some of the earliest and most interesting ancient cultures. Ancient Egypt was one of the birthplaces of human civilization. The Egyptians of pyramid fame were only one of the many cultures to make this arid land their home.

The Berbers: People of the Sand

A people called the Berbers settled along the western part of coastal North Africa, especially in the Atlas Mountain region. Most Berbers were traditionally small farmers and herders, although some groups, like the Tuaregs of the Sahara, were nomadic. The Berbers were also great traders and their trans-Saharan caravans played a major role over the centuries in linking the Mediterranean world with West Africa and its supplies of gold and slaves; some of their descendents continue traditional lives in the area.

Geographically Speaking

The Mediterranean Sea has always been one of the world's most important commercial and cultural crossroads. Its name in Latin means "middle of the earth," and that's exactly what it was in the ancient world—the place where the three known continents came together. Control of Mediterranean shipping lanes was a strategic goal of all the great powers that developed along its shores. And whoever controlled North Africa was well on the way to dominating the Mediterranean.

After the discovery of the Americas, Europe's commercial focus shifted to the Atlantic. But with the opening of Suez Canal in Egypt in 1856, the Mediterranean regained much of its importance as a trade link between the Atlantic and Indian Oceans.

The Arrival of Islam

Within a century of the death of Mohammed in 632 C.E., Muslim armies from Arabia had conquered Egypt and the rest of coastal North Africa. Although the Berbers of the Maghreb resisted at first, by 709 the Arab conquest of North Africa was complete.

Two years later, the Berbers even joined the Arabs in the conquest of much of Spain, where the invaders became known as Moors. The conquered populations of North Africa were mainly Christians, but over time most embraced Islam.

The Colonial Era

Beginning in the early sixteenth century, the Ottoman Turks added Egypt and most of coastal North Africa to their growing Mediterranean Empire. In 1882, Egypt became a de facto protectorate of Great Britain. Sudan, conquered by Egypt in 1874, was ruled jointly by Egypt and Britain from 1898 to 1955.

Like everywhere else in the world, the end of World War II signaled the beginning of the end of the colonial system in Saharan Africa. Throughout the region, the post-independence era has been marked by instability and turmoil, and in the case of the Sahel and the Horn of Africa, extreme poverty as well.

The Maghreb

Maghreb is the Arabic word for Northwest Africa. Technically, it refers only to the countries that share the Atlas Mountains—Morocco, Algeria, and Tunisia—but Libya is usually included in the group. (To be really technical about it, Maghreb refers only to the area between the Mediterranean and the high ranges of the Atlas Mountains.) In general, the terrain of the three Atlas Mountains countries consists of coastal plains, plateaus, and mountains in the north, and desert in the south. Libya is mostly flat arid plains and desert. Moving south from the Mediterranean Sea (which all four countries border) or east from the Atlantic (which only Morocco borders), the amount of annual rainfall plummets from 30 to 40 inches to less than 5 inches. Not surprisingly, 95 percent of the population lives in the coastal plains.

The Maghreb.

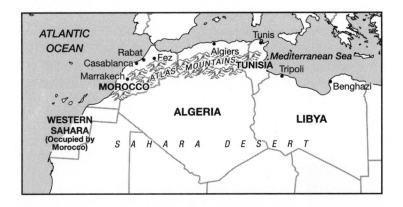

Morocco

Morocco is a relatively stable and open constitutional monarchy that has maintained good relations with the West and a moderate stance in the Arab-Israeli conflict. Morocco has come under international fire, however, for its de facto annexation of Western Sahara (formerly Spanish Sahara). In May 2003, a terrorist bomb attack, believed to be the work of al-Qaeda, killed 24 people in Casablanca.

Agriculture is still a key sector of Morocco's economy. Other industries include mining (phosphates), textiles, food processing, and tourism, which has taken a hit recently due to the global slowdown and terrorist threats.

Morocco has a population of over 33 million. Its capital is Rabat, with 1.6 million people; its largest city is Casablanca, with over 3 million, and the setting for a famous film by the same name. Morocco's tourist treasures include the trio of magnificent medieval royal cities: Fez, Meknes, and Marrakech.

Algeria

Algeria finally attained its independence from France in 1962 after eight years of bitter and violent war. The country's post-independence history has been even more violent. Since 1965, Algeria has been run essentially as a military dictatorship, despite the façade of more or less regular elections. When an Islamic fundamentalist party won the first stage of parliamentary elections in 1991, the military intervened to overturn the result, triggering a bloody civil war in which, according to some estimates, 100,000 people have been killed. Although an amnesty program started in 2000 has ended the worst of the violence, Algeria is still plagued by regular terrorist attacks, rising unrest among the Berber minority, and massive unemployment.

At least Algeria is not destitute. It has plenty of oil and even more natural gas (it's the world's second largest gas exporter). Of the country's 32 million people, nearly 4 million live in the metropolitan area of Algiers, the capital.

Terra-Trivia

Algeria is famous for the sirocco, a hot, dry, dust-filled wind that blows north from the Sahara Desert across the Mediterranean to Italy and its vicinity. Siroccos are most common in the spring.

Tunisia

Much of the western Mediterranean was once ruled from the great city-state of Carthage, whose ruins are located near Tunisia's capital, Tunis (population 700,000). Since gaining its independence from France in 1956, Tunisia has had only two leaders, both of whom have been moderate and basically pro-Western in foreign policy and repressive domestically. Islamist opposition parties are severely restricted, as are personal freedoms in general. On the other hand, the Tunisian government has been very progressive on women's issues and has done much to improve the social and legal standing of Tunisian women.

Tunisia has a relatively stable and diversified economy based on petroleum, mining (phosphates and iron ore), and agriculture. Its Mediterranean beaches and rich collection of ancient Carthaginian and Roman ruins and art are major tourist attractions. Of Tunisia's 10 million people, only 6 percent live below the poverty line, a very low figure for this region.

Libya

Libya became a monarchy after gaining its independence from Italy in 1951. The discovery of oil in 1959 turned the poverty-stricken county into an oil-rich nation, although much of its population has yet to benefit. In 1968, Colonel Muammar al-Qaddafi overthrew the king and established a radical, anti-Western, extremely anti-Israel, Islamic socialist regime. During the 1980s, Qaddafi diverted oil revenues to fund radical causes and terrorist attacks, most notoriously the downing of Pan Am flight 103 over Lockerbie, Scotland, in 1988. The United Nations imposed sanctions on Libya in 1992 that severely affected its economy. Those sanctions were not lifted until 2003, when Libya finally admitted its responsibility for the Lockerbie disaster and reached a financial settlement with the victims' families.

Libya's economy is all about oil, which accounts for almost all of its export income. The combination of oil revenues and a small population gives Libya one of the highest per capita gross domestic products (GDP) in Africa. Its unemployment rate, however, is a whopping 30 percent. Libya has a population of more than 5.5 million people, 1.6 million of whom live in the capital, Tripoli.

Egypt and Its Lifeline, the Nile River

Egypt is arguably the leading nation in the Arab world. Among the many reasons are its long and glorious history, both ancient and Islamic; its rich culture; and its strategic location at the junction of Asia and Africa. Egypt's population of more than 78 million is the largest in the Arab world. Its capital, Cairo, with 8 million people (nearly double that in its metropolitan area), is the largest city not only in the Arab world, but in all of Africa.

The Land, the River, and the People

Egypt is bordered on the west by Libya, on the north by the Mediterranean, on the east by Israel and the Red Sea, and on the south by Sudan. Egypt is basically one big sandy plateau, dotted by a few oases in the west, and divided into two unequal parts by its famous lifeline, the Nile River and its delta, basically desert with a narrow ribbon of Nile. The Nile enters Egypt from Sudan in the south and flows north for about 960 miles to the Mediterranean Sea. At the southern border with Sudan the Nile feeds Lake Nasser, formed by the Aswan High Dam. One of the world's largest artificial lakes, Lake Nasser is 300 miles long and 10 miles wide; about two thirds of it is in Egypt. Just north of Cairo, the Nile fans out to form its delta, which is about 155 miles wide when it reaches the Mediterranean Sea.

About 99 percent of Egypt's people live on 4 percent of the total land area—almost within sight of the Nile or its delta. The vast majority of the population is made up of ethnic Egyptians, Arabs, and Berbers, but in this highly cosmopolitan society there are also pockets of Greeks, Nubians, Armenians, and Europeans (mainly French and Italian). More than 90 percent of Egyptians are Sunni Muslims; the rest are mostly Coptic Christians. (The Coptic Church is the ancient native Egyptian Christian church; it is independent of both Orthodoxy and Catholicism.) Arabic is the national and official language; most educated Egyptians also speak English or French.

Post-Colonial Egypt

At least on paper, Egypt became an independent monarchy in 1922, but British domination of its political life didn't fully end until 1952, when an army group led by General Gamal Abdel Nasser overthrew the king and established an Arab socialist republic. Nasser nationalized the Suez Canal in 1956, triggering an unsuccessful invasion by Britain, France, and Israel. His aggressively anti-Israel policies led to the disastrous 1967 Arab-Israeli War, in which Egypt lost the Sinai Peninsula and Gaza Strip and much of its armed forces.

Nasser's successor, Anwar el-Sadat, launched the Yom Kippur War against Israel in 1973. Despite some early successes, the Arab side was again defeated by Israel. In 1978, U.S. president Jimmy Carter brokered the Camp David accords between Israel and

Egypt, and the following year Sadat became the first Arab leader to sign a peace treaty with Israel. In October 1981, Sadat was assassinated by Islamic extremists opposed to the treaty. Sadat's successor, Hosni Mubarak, has since kept the peace with Israel, although it is an increasingly cold peace as the Israeli-Palestinian conflict has escalated in recent years. Although Mubarak has introduced a number of economic reforms, freedom of political expression is still restricted.

Egypt's main economic resource is the Nile and its fertile valley. It also has oil and natural gas, phosphates, iron ore, and hydroelectric power from the Aswan Dam. The threat of terrorist attacks at home and a post-September 11, 2001, decline in international travel and shipping have had a negative effect on two of Egypt's main income earners, tourism and Suez Canal revenues.

The Sahel, or Transition Zone, Countries

South of the Sahara is the Sahel, the semiarid transition zone where the alternately expanding and shrinking desert turns into the grassy plains known as savannas. This area is also a transition zone from Arab to black African and from Islam to Christianity and indigenous religions.

The Sahel comprises five countries located in a horizontal band just below the Maghreb and Egypt. All are primarily desert or flat plain, with mountains in Northwestern Chad and East-central Sudan. The other main physical features of the subregion are the Nile River and its two main tributaries, the Blue Nile and the White Nile, in Sudan; Lake Chad, shared by mainly by Chad and Cameroon; and in the west, the great Niger River, which flows through Southeastern Niger and Southern Mali. The only two Sahelian countries that aren't landlocked are Mauritania and Sudan.

The Sahel.

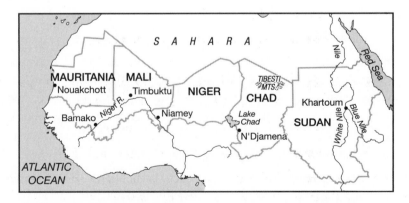

Mauritania, Mali, Niger, and Chad all gained their independence from France in 1960. Sudan's independence from Britain and Egypt was proclaimed in 1956. In the years since independence, all five countries have known authoritarian rule, political and ethnic strife, and extreme poverty. With little arable land and, with some exceptions, few natural resources, these are among the poorest countries in the world. Soil erosion and desertification (protracted drought, overgrazing, and agricultural use are turning the land into a desert) are major environmental issues throughout the region.

The Islamic Republic of Mauritania

Although Mauritania has been holding relatively open, multiparty elections since 1992, its government is still largely controlled by one party. In 1976, Mauritania took over the southern third of Spanish Sahara (today known as Western Sahara and controlled by Morocco), but withdrew in 1979 under mounting pressure from indigenous guerrillas. Ethnic tensions between the majority Maur (mixed Arab-Berber) and minority black populations are a persistent social and political problem. A military coup in 2005 ended President Taya's 21-year rule of the country. In terms of per capita gross domestic product, Mauritania is the most prosperous of the Sahelian countries (though that does not say much in this desperately poor region). Subsistence farming, fishing, and iron ore mining are the bases of its economy. Offshore oil reserves have recently been discovered. Mauritania has a population of around 3 million people, almost all of them Muslim. The capital, Nouakchott, has more than 800,000 people.

Mali

From the twelfth through the sixteenth centuries C.E., great trading empires flourished in what is now Mali. The fabled Malian cities of Timbuktu and Djenne grew vastly rich from the trans-Saharan caravan trade in gold, salt, and slaves and became major centers of Islamic culture. That period of greatness is long past. Despite recent political and economic reforms, Mali is desperately poor. About 80 percent of its workers live by farming and fishing in the only place where it's possible, in the southern region watered by the Niger and Senegal rivers. The vast majority of Mali's 13.5 million people are black Africans (mainly Mande); about 10 percent are Tuareg and Maur. The religious breakdown is Islam, 80 percent; indigenous beliefs, 18 percent. French is the official language; the majority African language is Bambara. The metropolitan area of Mali's capital, Bamako, is home to 1.3 million people.

Niger

The drop in world demand for uranium in the 1980s ended Niger's brief brush with prosperity. Since independence, Niger has experienced a succession of military governments, coups, assassinations, and ethnic unrest. In 1995, a longstanding conflict between nomadic Arab-Berber Tuaregs of the north and Niger's majority black population of the south ended in a cease-fire and later a peace accord. After two more coups in the late 1990s, democratic presidential elections were held in 1999. Niger is the poorest of the Sahelian countries, with, at last count, 63 percent of its population living below the poverty line. Subsistence farming, herding, and uranium mining are Niger's main economic activities. About 80 percent of Niger's nearly 14 million people are Muslim; the rest practice indigenous religions or Christianity. French is the official language; Hausa and Djerma are the main African languages. Niger's capital is Niamey, with 748,000 people.

Chad

Chad's post-independence story has been particularly violent and chaotic, filled with coups, military regimes, ethnic civil wars, and foreign interventions. The primary fault line is between the Muslim (50 percent of the population) north and east and the Christian or animist south, the traditional center of power. (The capital, N'Djamena, is in the south.) Libya's Muammar al-Qaddafi intervened throughout the 1970s and 1980s, either supporting Chadian factions or invading outright. Although Chad's current leader, in power since 1990, has attempted some political reforms, violence continues to flare in the troubled north.

Part of the problem in Chad is that its population of almost 9 million includes some 200 distinct ethnic groups. The economy is based on subsistence farming and herding, plus foreign aid. Oil, which the country started pumping in 2003, is one bright spot in Chad's economy.

Sudan

With an area of 968,000 square miles, Sudan is Africa's largest country; it also has a substantial population of 37 million. Geographically, Sudan's northern third is desert, its middle third is steppes (dry, level, treeless plains) and low mountains, and its southern third, called the Sudd, consists of vast swamplands. All the Nile tributaries flow through Southern Sudan, with the two main ones, the Blue Nile and the White Nile, meeting near the capital, Khartoum (metropolitan area population: 5.7 million), to form the Nile.

In the nearly 50 years it has been independent, Sudan has been free of civil war for only 10, from 1972 to 1982. Its governments have alternated between shaky parliamentary coalitions and military dictatorships. In Sudan, the north/south divide pits the dominant Arab Muslim population of the north against the animist or Christian black Sudanese of the south. In 1983, the government, under the increasing influence of Islamic fundamentalist groups, incorporated Shari'a (Islamic law) in the country's penal code. This action rekindled a civil war that over the last 20 years is thought to have cost more than 2 million people their lives. In addition, Sudan has come under international fire for harboring and supporting Islamic fundamentalist terrorist groups and for permitting the widespread enslavement of southern blacks.

Since July 2003, a full-scale, government-sponsored military operation known as the Janjaweed has attempted to annihilate the African tribes in the region with the support of Arab militias. In the Darfur region, attacks on civilians are an almost daily occurrence. Men, women, and children are tortured and killed, and villages burned to the ground. As many as 450,000 people are believed dead and more than 2.5 million are displaced. The United Nations and other outside powers have done little to help the suffering.

Sudan's economy is still overwhelmingly agricultural and poor. But since the mid-1990s, the government has implemented significant economic reforms that have helped fuel economic growth. In 1999, Sudan began exporting oil, which should help sustain growth over the coming years.

The Horn of Africa

East of Sudan the Sahara Desert meets the high Ethiopian Plateau. This marks the beginning of the Horn of Africa, a huge eastern outcropping that juts up and out from the continent into the Indian Ocean.

The terrain of the Horn of Africa is dominated by Ethiopia's high central plateau, which is split diagonally by the Great Rift Valley, a fault system that starts in Northern Syria, cuts through the Jordan Valley and the Red Sea, and in Africa, runs all the way down to Mozambique. In Ethiopia, the Rift Valley area is dotted by small volcanoes and hot springs in the north and by large lakes in the south. The central plateau is mountainous and has a generally temperate climate. It descends steeply to the hot, arid coastal plains of Somalia and the much narrower one of Eritrea. The Blue Nile, one of the Nile's main tributaries, rises in Lake Tana in Northwestern Ethiopia.

Eritrea, Ethiopia, and the western leg of Somalia are all former Italian colonies. Djibouti was once a French colony; the northwestern part of Somalia was formerly British. With the exception of Djibouti, the nations of the Horn of Africa have the unhappy distinction of being among the 10 poorest in the world in terms of per capita GDP.

Eritrea

After the Italians were driven out of the Horn of Africa in 1941, Eritrea was administered by Britain until 1952, when it joined Ethiopia in a federation. In 1962, Ethiopia made Eritrea a province, triggering a disastrous 30-year civil war. Eritrea finally prevailed in 1991 and became independent two years later. In 1998, a vicious and costly border war broke out between the two neighbors. Although hostilities ended in late 2000 and United Nation peacekeeping forces are in place, tensions are still high between the two countries.

Eritrea's 4.5 million people comprise nine different ethnic groups; Tigrinya and Arabic are the most common languages, although English is widely spoken. The population is almost evenly divided between Christians and Muslims. Nearly 900,000 people live in the capital, Asmara, and its surrounding area. Eritrea's war-bankrupted economy is based on subsistence farming.

The Horn of Africa.

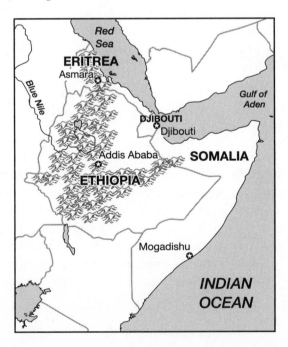

Ethiopia

Ethiopia is Africa's second oldest state after Egypt. According to legend, its first emperor was Menelik I, the son of Solomon and the Queen of Sheba. Except for the relatively brief period of Italian occupation from 1936 to 1941, Ethiopia has always been independent. In 1974, its last emperor Haile Selassie was deposed in a coup. The coup leaders set up a socialist military dictatorship, which became a full-fledged communist regime after Lt. Col. Mengistu Haile Mariam seized power in 1977. Supported by the Soviet Union, the Mariam government fought a civil war with Eritrean separatists and border wars with Somalia, while engaging in a reign of terror at home. After Mariam was forced into exile in 1991, Ethiopia's government recognized Eritrea's independence and has made efforts to defuse ethnic tensions and establish a multiparty democracy in the long-troubled nation.

Ethiopia has more than 80 ethnic groups, although three of them—the Oromo, Amhara, and Tigrean—account for about 75 percent of the country's 75 million people. About 2.9 million people live in the capital city of Addis Ababa. More than 70 languages are spoken in Ethiopia. Amharic is the official language; English is the most widely spoken foreign language. Nearly half the people are Muslim, more than a third are Ethiopian Orthodox Christians, and about 15 percent practice indigenous African faiths.

Coffee and foreign aid are the mainstays of Ethiopia's impoverished agricultural economy. The combination of war and recurring droughts over the last decades not only left Ethiopia's economy in shambles, but it also produced a series of horrifying famines.

Djibouti

Tiny Djibouti gained its independence from France in 1977. In 1991, mounting tensions between the dominant ethnic group, the Somali Issas, and the minority Afars led to the outbreak of a civil war that didn't fully end until 2001. The war, repeated droughts, and the influx of refugees from Ethiopia and Somalia have wrecked Djibouti's economy, although it remains the most prosperous in the Horn of Africa.

Djibouti has a population of nearly 800,000 people, about half of whom live in the capital of the same name. The official languages are French and Arabic; the predominant religion is Islam. Djibouti's economy is all about its strategic location at the junction of the Red Sea and Gulf of Aden, not far from the Arabian oilfields. Its port and railroad line are the country's main source of income and employment.

Somalia

East of Djibouti and Ethiopia and bounded on the north by the Gulf of Aden and on the east and south by the Indian Ocean is Somalia, a country of more than 8 million people, most of them Muslim Somalis. The former colonies of British and Italian Somaliland became the Republic of Somalia in 1960. The new country waged a costly territorial war with Ethiopia in the 1970s and faced mounting insurgencies in the 1980s. In 1990, Somalia began to disintegrate into hostile "ministates" controlled by rival warlords. War, famine, and mounting casualties led the United States to send in troops in 1992 to facilitate the delivery of food and assistance. U.S. forces withdrew in 1994 after the notorious "Black Hawk Down" incident in which the bodies of American soldiers were dragged through the streets of the capital, Mogadishu. As of mid-2003, Somalia has no national government and is a source of increasing concern as a likely base for al-Qaeda-linked terrorists. In December 2006, Ethiopia (backed by the United States) invaded Somalia to re-establish a democratic government after it was threatened by Islamic fundamentalists.

The Least You Need to Know

- The Sahara Desert, the largest in the world, is the dominant physical feature of Northern Africa.

- The Nile River, the longest in the world, is the source of life and civilization in an arid land.

- Ancient Egypt was one of the world's earliest civilizations.

- Islam is the predominant religion of Saharan Africa and acts as a unifying cultural link throughout the area.

- The nations of the Sahel and the Horn of Africa have had very troubled modern histories and are among the poorest countries in the world.

Chapter 21

Sub-Saharan Africa: Cradle of Humankind

In This Chapter

◆ The birthplace of humankind

◆ Ancient kingdoms and colonial masters

◆ Sub-Saharan Africa today and post-apartheid South Africa

About 70 percent of the more than 800 million Africans live in the 40 or so countries south of Saharan Africa.

Geographically, the sub-Saharan region is mainly one vast plateau covered with savannas (grasslands) in the north, tropical rain forests in the central equatorial belt, and more grasslands and deserts (the Kalahari and the Namib) in the south. Most of the region's lakes and mountains are located along the Great Rift Valley that cuts through Eastern Africa from Ethiopia to Mozambique. Sub-Saharan Africa's two great river basins are the Niger Basin in the southern part of the continent's western bulge and the Congo Basin in equatorial west-central Africa.

Sub-Saharan Africa is a region of stunning contrasts. The region is vastly rich in natural resources: petroleum, natural gas, gold, silver, diamonds, copper, uranium, tin, zinc, coal, iron—if it's a metal or a mineral, sub-Saharan Africa probably has it. And yet the reality of sub-Saharan Africa today is beyond grim. Along with the Sahel, sub-Saharan Africa is the least developed region on Earth, with subsistence agriculture still the leading economic activity in most of the region. Of the 25 nations with the lowest per capita GDP, 19 are in the Sahel and sub-Saharan Africa.

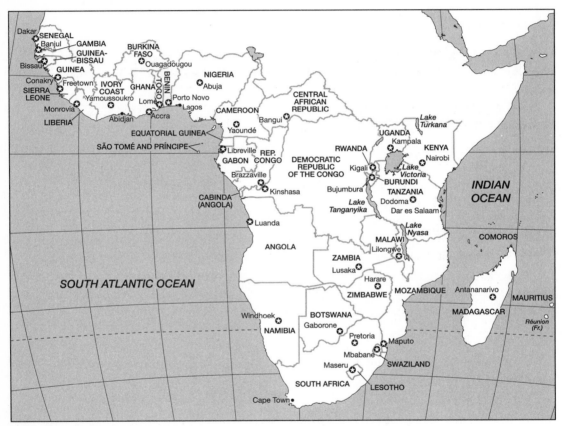

Sub-Saharan Africa.

Ancient Roots

Humans have lived longer in Africa than in any other continent. Recent fossil finds point to Ethiopia and East Africa as the likely birthplace of our species.

Although North Africa gave rise to great civilizations, sub-Saharan Africa remained isolated from developments in the Mediterranean world for thousands of years. By 1000 C.E., Muslim traders from the north were crossing the Sahara to trade salt for gold and slaves from West Africa and in the process brought Islam with them. Many great Bantu African kingdoms developed, including Ghana, Mali, Songhai, Benin, Great Zimbabwe, and the Kingdom of the Kongo.

The Rise of the Slave Trade

Beginning with the Portuguese in 1419, Europeans started to bypass the Arab slavers and aggressively exploited the region for slaves. This human trafficking went into high gear with the colonization of the "New World" and the need for cheap labor in the Americas.

Slavery ravaged the interior of Africa, destroying families, entire villages, and even cultures. It also caused untold misery and suffering to the slaves themselves and their descendants for centuries to follow. Although European traders introduced New World foods to Africa, such as corn and cassava, that would become important crops on the continent, they also introduced guns, which helped to further destabilize African societies.

Terra-Trivia

Liberia, Africa's first republic, was founded by the American Colonization Society in 1822 as a private colony (not part of the United States) for the settlement of freed slaves. Within 40 years, some 12,000 former slaves had settled what was then called Monrovia, after U.S. president James Monroe. In 1847, Monrovia became the Republic of Liberia, with the city of Monrovia as its capital.

The European Takeover

Beginning in the 1800s, European powers divided Africa up among themselves and imposed their rule on the varied African peoples. Their single goal was to enrich their European homelands without any concern for the people of the region. They imposed artificial borders without any consideration of tribe, clan, language, or other native cultural identities, imposed their European languages on the people, and turned Africa into a virtual "slave continent."

Independent Africa: Instability and Turmoil

At the end of World War II, even the victorious European powers were too devastated to hold on to their empires. By the late 1960s, almost all the sub-Saharan nations had gained their independence. During this postcolonial era, sub-Saharan Africa has known much more than its share of violence, warfare, epidemic disease, environmental destruction, famine, and even genocide. From the 1960s through the 1980s, much of the region became a surrogate battleground for the cold-war struggles between the Soviet Union and the United States. Military coups, countercoups, and brutal military regimes have been the norm across the region. Nationalist struggles and inter-tribal conflicts nearly tore apart Nigeria and the Republic of the Congo.

An Overview of Sub-Saharan Africa

Sub-Saharan Africa has so many countries that a description of each one is beyond the scope of this book. This section divides the region into five subregions and describes the highlights of each.

Western Africa

This subregion encompasses the southern part of the Western African bulge, from Senegal to Nigeria. All the countries are coastal except for Burkina Faso. Because the area is tropical, it's hot year-round. Rainfall is heavy in the south and diminishes to the north. Tropical rain forests in the south give way to savanna in the northern half of the subregion.

The dominant physical feature of the area is the Niger River. The world's thirteenth longest river (2,600 miles long) rises in Guinea, flows northeast out of the subregion into the Sahelian countries of Mali and Niger and then back down through Nigeria where it empties into the Gulf of Guinea. Lake Volta, one of the world's largest artificial lakes, is located in Ghana, and the northeastern corner of Nigeria borders on Lake Chad.

Western Africa has about 222 million people. Its largest country is Nigeria, with around 133.5 million (in fact, Nigeria is the most populous country in all of Africa). The cities along the southern coast of Western Africa form one gigantic megalopolis, from Abidjan, the administrative capital of Côte d'Ivoire (Ivory Coast), with over 4 million people; to Accra, capital of Ghana, with nearly 3 million people; and Nigeria's largest city, Lagos, with 9.5 million. To the north, Senegal's capital, Dakar, is also a major metropolis, with a population of nearly 2.5 million. Despite its number of large cities, Western Africa is still heavily rural. Even in oil-rich Nigeria, nearly 70 percent of the workforce is engaged in agriculture.

The tribal, ethnic, and linguistic makeup of the subregion is extremely complex. Nigeria alone has more than 250 ethnic groups. Senegal and Gambia are more than 90 percent Muslim; in the rest of the nations of the subregion, varying percentages of the population practice either Islam, Christianity, or indigenous faiths.

In size and population, Nigeria is the cornerstone of Western Africa. It is also potentially one of Africa's wealthiest nations thanks to its oil reserves. An oil boom in the 1970s left Nigeria with the thirty-third highest per capita gross domestic product in the world. Today, thanks to corruption, mismanagement, and civil strife, Nigeria is one of the poorest countries in the world. Its stability is once again seriously threatened by mounting tensions between Christians and Muslims. The most prosperous countries in the subregion in terms of per capita gross domestic product are Ghana and then Guinea; the poorest are Sierra Leone and Guinea-Bissau, which are among the poorest countries in the world.

Central Africa

Central Africa lies astride the equator in the west-central part of the continent. From north to south along the Atlantic coast are Cameroon, Equatorial Guinea, Gabon, Congo, a tiny separated bit of Angola called Cabinda, and the Democratic Republic of Congo's narrow outlet to the ocean. Inland, north of the Democratic Republic of Congo, is the Central African Republic. Off the western shore is the island nation of São Tomé and Príncipe.

Rainfall is ample throughout this tropical subregion, although it's heavier along the coast and on the mountain slopes. West Cameroon's slopes get as much as 400 inches of annual rainfall! Two of the subregion's major rivers (it has hundreds) separate the Central African Republic from the Democratic Republic of the Congo: the 2,880-mile-long Congo (the world's eighth longest) and its major tributary, the 660-mile-long Ubangi.

Northern Cameroon and most of Central African Republic are covered by savanna and shrub growth. The rest of the subregion is mostly tropical rain forest. The heart of the subregion is the Congo Basin, shared by the two Congos. Most of the basin is surrounded by plateaus and mountains, the tallest being the Mitumba Range along the Democratic Republic of the Congo's eastern border.

Central Africa is relatively thinly peopled. Only 40 percent of the population is urban. The largest city is Kinshasa, the capital of the Democratic Republic of the Congo. This huge and troubled country in the heart of Africa is the giant of the subregion, with more than half of the land and people.

Although Bantu-based languages prevail in most of the subregion, the legacy of France and (in the case of the Democratic Republic of the Congo) Belgium is evident in the fact that French is the official language of the two Congos, the Central African Republic, and Gabon. French and English are the official languages of Cameroon, while Portuguese is spoken in São Tomé and Príncipe. Equatorial Guinea is the only country in Africa whose official language is Spanish.

Most people in Central Africa farm, fish, herd, log, or mine for a living. The subregion includes the most prosperous sub-Saharan country outside of South Africa as well as one of the poorest. A combination of offshore oil, a small population, a history of stable (if authoritarian) government, and more recently, a freer and more open political system explains the relative prosperity of Gabon today.

On the opposite side of the coin, a grim history of colonial exploitation, secessionist movements, the long and disastrous dictatorship of General Mobutu (1965 to 1997), civil wars, and ongoing ethnic conflicts both at home and with its neighbors has left the Democratic Republic of the Congo devastated and impoverished. Even though the country possesses vast amounts of natural and mineral resources—oil, gold, silver, diamonds, uranium, zinc, you name it—the DRC's per capita GDP at $700 ranks among the five poorest countries in the world.

Eastern Africa

The subregion of Eastern Africa straddles the equator in the east-central part of the continent. Kenya and Tanzania are on the Indian Ocean coast; Uganda, Rwanda, and Burundi lie just to the west. The climate is tropical year-round but cooler in higher elevations. Although Uganda, Rwanda, and Burundi have ample rain, Tanzania is generally drier and Kenya considerably drier still. From west to east, this highland region ranges from light tropical forest vegetation to savanna lands.

The Nile rises in Burundi and Rwanda, but it's small and inconsequential in this area. A string of mostly elongated lakes marks the entire western edge of the subregion. These include Lake Tanganyika, the seventh largest lake in the world, and Lake Malawi, the tenth largest. These lakes and the surrounding mountains are all part of the Great Rift Valley system.

Eastern Africa is also home to two of the continent's great scenic wonders. Lake Victoria, the world's second-largest freshwater lake, lies on the equator where the boundaries of Uganda, Kenya, and Tanzania meet. Although Eastern Africa is not as mountainous as Ethiopia to the north, Northern Tanzania does contain Africa's highest mountain, the snow-capped Mount Kilimanjaro, at 19,340 feet. Unfortunately in 2005, Mt. Kilimanjaro was found to be snow free for the first time in 11,000 years, and by 2020, the snow cap is expected to have completely vanished.

The fertile Lake Victoria basin is one of Africa's major population centers. Almost half of Eastern Africa's 106 million people live in the parts of Uganda, Kenya, and Tanzania surrounding the lake as well as in tiny Burundi and Rwanda. The population density here exceeds 500 people per square mile, compared to 65 for all of sub-Saharan Africa. With about 2.4 million people each, the largest cities in Eastern Africa are Kenya's capital, Nairobi (3–4 million), and Tanzania's administrative capital, Dar es Salaam (2.5 million). (Tanzania has also moved its capital inland to the small city of Dodoma.) The sub-region's next largest cities are Uganda's capital, Kampala (1.2 million), and Mombasa (665,000) in Kenya.

Despite its large urban population, Eastern Africa is still overwhelmingly rural. All of the countries in this subregion cluster at the bottom of the poverty chart. With a per capita GDP of $1,200, Uganda is the best off of the group. It has achieved a measure of political stability recently and it has an abundance of fertile soil and regular rain to support its primarily agricultural economy (coffee is the main export). Uganda has also been one of the most successful countries in Africa in fighting AIDS, the lethal disease that is ravaging much of sub-Saharan Africa.

Kenya is the financial and business hub of Eastern Africa, but its economy is plagued by deep-seated corruption and low prices for its main exports: coffee, tea, and fruits and vegetables. Tanzania had its first multiparty elections in 1995 and its heavily agricultural economy has been growing in recent years due to a pickup in the mining sector (diamonds and gold). Nonetheless, Tanzania (per capita GDP at $700) remains one of the poorest countries in the world. In Eastern Africa, only Burundi is as poor. The gruesome events in the recent histories of Burundi (1972 genocide, 100,000–150,000 killed; 1993 genocide, 25,000 killed) and Rwanda (1994 genocide resulted in up to 1 million deaths) have deeply scarred the region.

Southern Africa and Madagascar

Southern Africa includes everything south of the Democratic Republic of the Congo and Tanzania. Heading south on the Atlantic coast on the west are Angola, Namibia, and South Africa (which surrounds Lesotho and Swaziland). Mozambique takes up the entire Indian Ocean coast on the east, north of South Africa. In the center are the landlocked countries of Malawi, Zambia, Zimbabwe, and Botswana. Madagascar, the world's fourth-largest island, lies off Mozambique's east coast, along with the island countries of Comoros and Mauritius and the tiny French island of Réunion. Because of its economic importance, South Africa is treated separately in the following section of this chapter.

Woodland forests dominate Angola, Zambia, Malawi, Northern Mozambique, and Eastern Madagascar. The Kalahari Desert covers Western Botswana and Eastern Namibia, and extends into South Africa. The Namib Desert runs all along the coast of Namibia into South Africa and north into Angola. Almost everywhere else, croplands and grazing lands prevail.

Although Northern Angola, Zambia, and Malawi are tropical, the rest of the region is temperate with mild winters and hot summers. The deserts are hotter and dry. The forest areas in the north get moderate amounts of rain; totals are lower elsewhere in the subregion.

The largest lake in the subregion is Lake Malawi (Lake Nyasa), between Malawi and Mozambique. The region has several sizable rivers, including the Zambezi (Africa's fourth longest), which forms the border between Zambia and Zimbabwe; the Limpopo, which forms the eastern part of the border between Botswana and South Africa; and the Orange-Vaal, which flows through central South Africa before forming the border between South Africa and Namibia. The subregion is also home to another of Africa's great natural wonders: the Zambezi River's Victoria Falls (the largest single sheet of water in the world, over 1 mile wide) on the eastern end of the border between Zambia and Zimbabwe.

In addition to the big cities of South Africa—Cape Town, Johannesburg, and Durban—Southern Africa's other large cities include Angola's capital, Luanda (4.5 million); Zimbabwe's capital, Harare (1.6 million–2.8 million in metro area); Mozambique's capital, Maputo (nearly 1 million), and Zambia's capital, Lusaka (just over 1 million). Except for South Africa, the countries in this area are thinly populated: They have 25 percent of sub-Saharan Africa's land but only about 13 percent of its people.

Excluding South Africa, Botswana is the wealthiest country in the subregion and one of Africa's rare economic and political success stories. Since gaining independence from Britain in 1966, Botswana has maintained a stable democracy (now Africa's oldest) and its diamond-mining industry has fueled healthy economic growth and the second highest per capita GDP in Africa ($10,700). (Tourism and agriculture are its other main industries.) Unemployment and poverty rates remain high, however, and even more troubling, Botswana has one of the highest HIV/AIDS infection rates in the world.

In the rest of the subregion, the economic picture is grimmer. Violent independence struggles, communist dictatorships, and years of civil wars left both Angola and Mozambique in tatters. Although Mozambique has achieved a good measure of political stability and even economic growth since 1992, its already impoverished economy was dealt a heavy blow by severe flooding in 1999 and 2000. Thanks to its oil and diamonds, Angola is somewhat richer than Mozambique, but in many respects it is even worse off. From 1975 to 2002 the country was locked in a civil war that is estimated to have cost 1.5 million lives. Although the peace seems to be holding, Angola is struggling with food scarcities as thousands of people who had fled the country during the war return home. Life expectancy in Angola (38.43), Swaziland (33.22), Botswana (33.87), Lesotho (34.47), Zimbabwe (37.82), Zambia (39.70), and Mozambique (40.32) are all among the lowest 10 countries in the world.

Copper-rich Zambia has been hard-hit by low copper prices, droughts, and the usual problems of political instability and corruption. Desperately poor Malawi has held multiparty elections in 1994 and 1999, but its economy is among the world's least developed (per capita GDP $600). With agriculture accounting for more than 90 percent of the workforce, Malawi barely survives on tobacco exports and foreign aid. A growing HIV/AIDS epidemic is taking a heavy toll here as well.

The destructive economic policies of Zimbabwe's renegade president, Robert Mugabe, the only ruler Zimbabwe has known since independence, have caused its white farmers to flee and have wrecked the potentially healthy economy of this mineral-rich country. Inflation in early 2003 was 238 percent, economic growth declined more than 12 percent in 2002, and an estimated 70 percent of Zimbabweans live below the poverty line.

Madagascar held its first multiparty elections in 1992, but a contested election in 2001 nearly split the country apart. Agriculture is the mainstay of this extremely poor country's economy.

Eye on the Environment

Madagascar is known as the "land of living fossils," as it is home to a huge number of wildlife species found nowhere else on Earth. It is especially famous for its lemurs (primitive tree-dwelling primates) and chameleons. Deforestation is a serious and growing threat to the rainforests that support Madagascar's unique animals and plants.

South Africa: Challenges in the Post-Apartheid Era

Gold- and mineral-rich South Africa is the continent's most economically developed and wealthiest country. It is the world's biggest producer of gold, platinum, and chromium, and a leading producer of diamonds and a host of other metals and minerals. It has a diversified economy, with a well-developed infrastructure, a sound fiscal system, and a major stock exchange (one of the world's 10 largest). That's the good news. The bad news is that South Africa's economy continues to be weighed down by the high poverty rates and lack of equal economic opportunity for its nonwhite citizens—lingering legacies of South Africa's now dismantled system of racial separation known as apartheid.

South Africa's terrain is dominated by a high central plateau surrounded by mountains on the east and south; the highest elevations are in the Drakensberg Mountains in the east. Most of the plateau is rolling grassland, or veld, called highveld (over 5,000 feet). To the west is the lower, semiarid middleveld, and between the mountains and the sea is the lowveld, bordered by a narrow coastal strip. The southern reaches of the Kalahari and Namib Deserts extend into South Africa from Botswana and Namibia. South Africa is the only truly temperate country in sub-Saharan Africa. Although temperatures vary by season and altitude, hot summers and mild winters typically prevail. Rainfall is moderate in the east but sparse in the west.

South Africa has a population of 47.4 million. About 75 percent are black, 16 percent are white, 10 percent are mixed race, and 2.6 percent are of Indian descent. There are 11 official languages: Afrikaans (the first language of about 60 percent of the white population and most mixed-race people), English, and nine indigenous languages. About 68 percent of the population are Christian and more than 28 percent practice animism and other indigenous faiths. There are also small percentages of

Hindus and Muslims. South Africa actually has three capitals: Pretoria, the executive and semi-official capital (population: 1.9 million); Cape Town, the legislative capital (population: 3 million); and Blomfontein, the judicial capital (with 645,000 people). Johannesburg is South Africa's largest city with 3.2 million and has the biggest metropolitan area population at around 7 million (because of phenomenal growth, it is expected to meld with surrounding satellites and Pretoria to form a megacity which by 2015 will be the twelfth largest in the world).

Although the apartheid regime has been successfully dismantled and black-majority rule is firmly in place, South Africa still struggles to bridge huge economic and racial divides within its population. Although whites still dominate the economy, blacks, often poorly paid and ill-treated, make up more than three quarters of the workforce. Despite the country's wealth, about 50 percent of South Africans live in poverty. South Africa also suffers from high crime rates and one of the world's worst HIV/AIDS crises. It is estimated that South Africa has more HIV-infected people (about 5 million) than any other country on Earth (over 30 percent of pregnant women are infected with HIV).

In the News: Africa's Health Crises

HIV/AIDS has been mentioned frequently in this chapter and for good reason. The disease today is the leading cause of death in sub-Saharan Africa. In 2005, 24.5 million people in the region were infected with HIV, 2 million died of AIDS in that year, and 12 million children were orphaned by their parents' AIDS death. In a region where overpopulation was once the problem, population growth rates have either slowed dramatically or are declining in some places because of mounting infant and adult death rates due to AIDS. Life expectancy rates are dropping throughout the region, many to below 40. (In relatively wealthy Botswana, life expectancy would have been 72 years had it not been for AIDS.)

The consequences of AIDS in Africa go beyond individual suffering and loss of life. An entire generation is being targeted just at the age when it should be at or near peak productivity. The effects on the economies and social fabrics of many sub-Saharan nations threaten to be catastrophic. As of 2001, there were an estimated 11 million AIDS orphans in the region, and that number is expected to climb to 20 million by 2010. South Africa is projected to lose 11 percent of its workforce by 2005; Zimbabwe, about 20 percent. Compounding the crisis is the fact that the medications that have made AIDS in the west more of a chronic disease than an instant killer are either unavailable in Africa or out of most people's reach.

The Least You Need to Know

- ◆ In the Middle Ages, sub-Saharan Africa gave rise to wealthy and sophisticated trading kingdoms.

- ◆ The African slave trade inflicted unspeakable misery and suffering on the peoples of sub-Saharan Africa.

- ◆ In an unprecedented burst of imperialism, European countries carved up Africa into colonies in the late nineteenth and early twentieth centuries.

- ◆ Although all the sub-Saharan African countries are now independent, only a few have attained political stability and most are very poor.

- ◆ AIDS is taking a catastrophic toll throughout sub-Saharan Africa.

The Middle East and Central Asia: Crucible of Faiths

In This Chapter

- ◆ Untangling a tangled region
- ◆ Islam's heartland: the Arabian peninsula
- ◆ Regional tinderbox: the Middle East core
- ◆ Ancient Seats of Empire: Turkey and Iran
- ◆ East-West Crossroads: Central Asian -stans

As of Fall 2001, the Middle East and Central Asia have become the main battlegrounds in the global war on terror. In the aftermath of the September 11, 2001, al-Qaeda-led terrorist attacks in New York City and Washington, D.C., the United States and its allies invaded Afghanistan in late 2001 and brought down the Taliban regime that had harbored al-Qaeda leader Osama bin Laden.

In March 2003, the United States targeted Iraq and toppled its murderous dictator, Saddam Hussein. In May 2003, a bombing in Riyadh, the capital of Saudi Arabia, killed 34 people and signaled that terrorism had arrived in the very heart of the Islamic world. The always-simmering conflict between Israelis and Palestinians over who will control what bits of the land that three religions call holy has erupted into a nightmare of hatred and violence and flared into open warfare with the Hezbollah provocation and subsequent Israeli invasion of Lebanon in July 2006.

Crossroads of Three Continents

We are actually dealing with three regions—the Middle East, Transcaucasia, and Central Asia—that are linked by a common religion, Islam, and a lot of shared history. But even defining the term "Middle East" isn't that easy. Europeans came up with it as a name for the region they viewed as being midway between themselves and East Asia (the "Far East"). The term usually refers to the lands of Southwestern Asia that lie west of Afghanistan and Pakistan—in other words, Iran and everything to the west (except for the Caucasus countries, which are not considered to be Middle Eastern). Egypt and even Libya are often included in the Middle East.

Depending on your definition, the Middle East either includes part of Africa (Egypt) or abuts it at the Israeli border. The region also includes a small part of Europe via Turkey, which straddles the Europe/Asia divide at the Bosporus and Dardanelles straits. This makes the Middle East the only region in the world that encompasses or at least abuts three continents.

Like Turkey, Iran is a Muslim country, but not an Arab one. So Iran's western border marks the end of the Arab Middle East, while its eastern border marks the beginning of Central Asia. The Central Asian region includes Afghanistan and all the former republics of the Soviet Union whose names end in -*stan* (meaning "place of").

To make this chapter a little more manageable, we've divided the dual Middle East and Central Asia region into five subregions:

◆ **The Arabian Peninsula:** Bahrain, Kuwait, Oman, Qatar, Saudi Arabia, United Arab Emirates, and Yemen

◆ **The Middle East core:** Israel, the West Bank and Gaza, Jordan, Lebanon, Syria, and Iraq

◆ **The Outer Middle East:** Turkey, Cyprus, and Iran

◆ **The Southern Caucasus:** Armenia, Azerbaijan, and Georgia

◆ **The Central Asian -stans:** Afghanistan, Kazakhstan, Kyrgyzstan, Tajikistan, Turkmenistan, and Uzbekistan

This general overview of each subregion should give you a better understanding of the issues and forces at play throughout the Middle East and Central Asia.

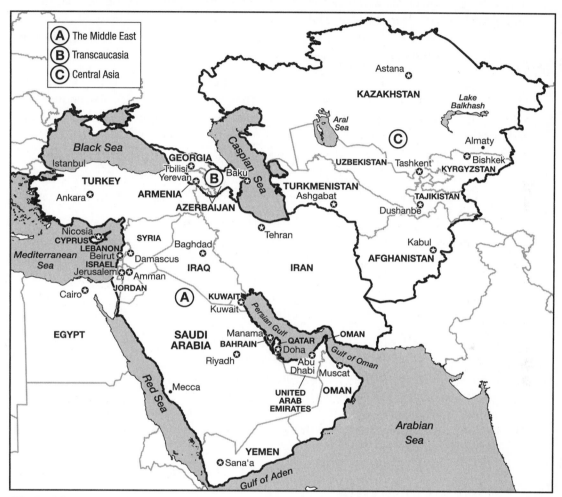

The Middle East and Central Asia.

The Middle East: Land of Faith, Oil, and Conflict

There are two key facts about the Middle East that are at the crux of its current turmoil. First, the region was the crucible for three of the world's great religions. Judaism, Christianity, and Islam all claim either spiritual or actual descent from Abraham and all hold the city of Jerusalem to be sacred. The other fact is oil. Petroleum was discovered in Persia (now Iran) in 1908 and in the Arabian peninsula in the 1930s. Today, about three fifths of the world's known oil reserves are in the Middle East, mainly in Saudi Arabia, Iraq, Iran, Kuwait, and the United Arab Emirates. Control of the land that is now the state of Israel has been a source of conflict among Christians, Muslims, and more recently Jews for centuries.

Even before there was oil, the Middle East was a much sought-after prize. Whoever controlled the Middle East controlled one of the world's great trade and conquest routes. Since the dawn of civilization, the region has been conquered and lost countless times.

Settled agriculture in the Middle East dates back to at least 8000 B.C.E. Around 3500 B.C.E. one of the world's first great urban civilizations, that of Sumer, emerged in the fertile Mesopotamian plain between the Tigris and Euphrates Rivers in what is now Iraq.

The region has seen one empire, and more conquerors, after another. The Philistines, Hebrews, Phoenicians, Greeks, Macedonians under Alexander the Great, the Romans, Persians, Turks, and countless others made this region's history complex. Along the way, Solomon built his Holy Temple, Jesus of Nazareth taught his message of peace, and Mohammed became the prophet of Islam.

In recent centuries, the power of the Byzantine Empire was replaced by the Ottoman Turkish Empire. After World War I, the Turks ceded control of the area to the British and French. The British Mandates in the area became present-day Iraq, Jordan, Israel, Kuwait, Qatar, United Arab Emirates, Oman, and South Yemen, while France granted independence to Syria and Lebanon. Much of this occurred without local consent and without regard for cultures, religion, or tribal identities, which is still a source of difficulty today.

The Arabian Peninsula

The Arabian peninsula is bordered on the north by Jordan and Iraq, on the east by the Persian Gulf and the Gulf of Oman, on the southeast by the Arabian Sea, on the south by the Gulf of Aden, and on the west by the Red Sea and the Gulf of Aqaba. Seven countries share the peninsula. The subregion's giant and namesake is Saudi Arabia (capital: Riyadh), birthplace of Islam and the world's biggest oil producer. Going clockwise from the northeast are the small Persian Gulf oil states: Kuwait (capital: Kuwait), whose invasion by Iraq in 1991 triggered the first Gulf War; tiny

Bahrain (Al-Manamah); Qatar (Doha); and the United Arab Emirates (Abu Dhabi). On the peninsula's southern end are Oman (Muscat) and Yemen (San'a).

The countries of the Arabian peninsula are mostly desert, complete with rolling sand dunes. Although the Red Sea marks the end of Africa and the beginning of Asia, the Arabian desert is actually an extension of Africa's great Sahara desert. Along the coast of the Red Sea and the Gulf of Aden, mountain chains rim the peninsula, with some peaks rising higher than 10,000 feet; there are also mountains in northern Oman. Most of the peninsula has a harsh, desert climate with huge temperature extremes. There are no rivers to speak of, but the mountainous southwestern part of the peninsula does get seasonal monsoon rains. Lack of arable land and a scarcity of fresh drinking water are major issues throughout the subregion.

The 50 million people of the peninsula are almost all ethnic Arabs, although there are also small South Asian and African minorities. Arabic is the main language of the peninsula; English is a fairly common second language. Some Indian languages are also spoken, mainly in areas with large numbers of guest workers hailing from South Asia.

In the homeland of the Prophet Mohammed, Islam is the nearly universal religion of the Arabian peninsula (the exceptions are essentially foreign workers). Mecca in Saudi Arabia is the birthplace of Mohammed and Islam's holiest site. One of the five basic obligations of every Muslim is to make a pilgrimage to Mecca, called the *Hajj*, at least once in a lifetime. Also in Saudi Arabia is the holy city of Al Madinah (Medina), where Mohammed died and was buried in 632 C.E.

With one exception, the countries of the Arabian peninsula are economically all about oil. The peninsula's Big Three oil producers (Saudi Arabia, Kuwait, and the United Arab Emirates) alone have more than 40 percent of the world's known oil reserves. The per capita gross domestic products (GDP) of the oil-producing nations of the peninsula are all very high, with Qatar and the United Arab Emirates leading the pack. In contrast, Yemen, which has had a turbulent modern history and began producing oil only in the mid-1990s, is one of the poorest countries not only on the peninsula but also in the world (per capita GDP $900).

But even in oil-rich Saudi Arabia there are economic clouds overhead. A burgeoning population, high unemployment, and overdependence on world oil prices are increasingly serious problems for the Saudi economy. A long-term decline in oil revenues has caused the Saudi government to cut back on many of the perks its citizens once took for granted.

Although on the surface the Arabian peninsula seems more stable than the rest of the Middle East, the underlying reality is disturbing. In general, the governments of the region are autocratic monarchies or emirates with close political and economic ties to the West. Their populations, however, are increasingly anti-American and fundamentalist Islamic. The 9/11 mastermind, Osama bin Laden, is the son of an extremely wealthy Saudi family, and 15 of the 9/11 suicide bombers were Saudi citizens. A major source of anger among conservative Saudis was the presence since the 1991 Gulf War of U.S. troops on Saudi Arabian soil. Although the United States withdrew its forces from the country in April 2003, the U.S. occupation of Iraq and the ongoing Palestinian-Israeli conflict will likely continue to fuel Islamic militancy on the Arabian peninsula.

The Middle East Core

One of the "cradles of civilization" and the wellspring of Judaism and Christianity, the Middle Eastern Core has been one of the world's most fought-over regions. Much of the contention surrounds the ancient city of Jerusalem, which is sacred to Jews, Christians, and Muslims alike. To Jews, it's the City of David, the ancient capital of Israel, the site of Solomon's and later Herod's Temple, and the holy Western Wall, the most sacred spot on earth for Jewish people. To Christians, it is the city where the crucifixion and resurrection of Jesus took place. Muslims regard Jerusalem as their third holiest city after Saudi Arabia's Mecca and Medina. The great Dome of the Rock, built on top of the remains of the Temple of Herod, enshrines the spot where Mohammed ascended to heaven. Today, both Israeli Jews and Palestinian Arabs claim Jerusalem as their capital.

The Middle East Core is bordered on the north by Turkey; on the east by Iran; on the south by a bit of the Persian Gulf, Kuwait, and Saudi Arabia; and on the east by Egypt and the Mediterranean. The subregion includes five countries and one entity whose status remains unresolved. The five nations are, clockwise from the north: Syria (capital: Damascus), Iraq (Baghdad), Jordan (Amman), Israel (Jerusalem), and Lebanon (Beirut). The status of the Palestinian territories of the West Bank (west of the Jordan River) and the Gaza Strip (the narrow coastal strip south and west of Israel) remains undecided and is currently being bitterly contested. Iraq's status is also uncertain. After the U.S. and British invasion in March 2003 overthrew Iraq's long-time dictator, Saddam Hussein, the country remains under mainly U.S. and British occupation and U.S. administration. A transfer of power to an elected government is currently shored up by a huge U.S. military force and has not brought much of a change to the Iraqi turmoil. In fact, the situation is basically a civil war with the majority Shiites dominating, but sectarian death squads murdering on both sides of the Islamic religious divide. The one bright spot is the Kurd areas of Northern Iraq, which are functioning almost autonomously and enjoy a basic level of prosperity and peace.

The Middle East core.

The Lay of the Holy Land

Syria, Lebanon, Israel, and the Gaza Strip all border the Mediterranean. Jordan and Iraq are almost landlocked except for Jordan's narrow outlet to the Gulf of Aqaba and Iraq's narrow but vital outlet to the Persian Gulf. Two deserts cover much of the subregion: the Negev in Southern Israel and the much larger Syrian Desert, which encompasses Southeastern Syria, Western Iraq, and a good part of Jordan. Western and Northern Syria and Northeastern Iraq are somewhat mountainous. Inland from the Mediterranean coast, the Great Rift Valley cuts through the subregion from Syria through the Jordan Valley before continuing down to Africa.

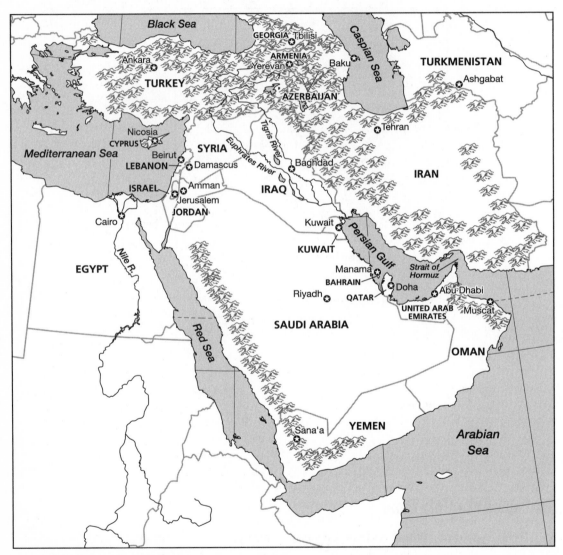

The Middle East—physical features.

The Jordan River of biblical fame is the lifeline of Jordan, Israel, and the Palestinian territories. (Not surprisingly, control of scarce water supplies is a big bone of contention in this part of the world.) But far and away the region's most important river system is located in Eastern Iraq. The Tigris and Euphrates Rivers both rise in Turkey and flow into Iraq (the Euphrates via Syria). The arc formed by Syria's Mediterranean coast, its part of the Euphrates valley, and the Tigris and Euphrates river system in Iraq make up the famous Fertile Crescent, which nurtured great civilizations and mighty empires for millennia.

The climate in much of the subregion is hot and very dry. The coastal areas, however, have a typically Mediterranean climate, with long, hot, dry summers, and short, cool, rainy winters.

The overwhelming majority of the 57 million or so people of the Middle East Core are Arabs. The main exception is Israel, which is about 76 percent Jewish. Israel's non-Jewish population is predominantly Palestinian Arab. There are also Armenian minorities in Syria, Lebanon, Israel, and a tiny one in Jordan. Syria also has a small Kurdish minority, while Iraq has a sizeable one. Nearly 20 percent of Iraqis are Kurds, living in what currently amounts to an autonomous state of their own in Northern Iraq.

Muslims live in every country of the Middle East Core and are the majority in most of them. Israel is the most notable exception. Only about 20 percent of its population is non-Jewish; about 15 percent are Muslim and 2 percent Christian. Although both Syria and Lebanon are more than 90 percent Arab, both have significant Christian minorities—10 percent in Syria and 30 percent in Lebanon. (Lesson: Not all Arabs are Muslim.) Iraq, on the other hand, is about 97 percent Muslim.

> **Terra-Trivia**
>
> Despite their current conflicts, Jews and Arabs are both Semitic peoples. The term *Semitic* refers to a group of peoples who speak (or spoke) related languages all thought to stem from a common Semitic tongue. In addition to Jews and Arabs, the following groups, all familiar from the Bible, were also Semites: the Assyrians, Babylonians, Canaanites, Phoenicians, and Aramaeans (who included the Hebrews, or Jews).

Israel, the Palestinian Territories, and Jordan

The escalating Palestinian-Israeli conflict, the global economic slowdown following the September 11, 2001, terrorist attacks, and the Iraq war and its aftermath have all badly hurt the economies of the subregion. Israel's economy, the most developed and technologically advanced in the Middle East Core, has been in a recession since the beginning of the second Palestinian *Intifada* ("uprising"). The reality or threat of suicide bombings in the buses and cafes of Israel's main cities, Tel Aviv and Jerusalem, have taken a heavy toll on the country, psychologically and economically. On the other side of the Israeli-Palestinian fence, Israeli crackdowns and military attacks in response to suicide bombings and rocket attacks from the Palestinian territories have destroyed much of the infrastructure in the Palestinian territories and Southern Lebanon and severely hampered freedom of movement and the ability to work. The result is that the Palestinian economy, fragile to start with, is in shambles. Foreign aid is all that is keeping it from collapsing—currently, the Gaza Strip has the lowest per capita GDP in the world. A further complication is the ongoing power struggle between the two largest Palestinian factions, Hamas and Fatah. Both militant groups have attacked and killed members of the other Palestinian party and Palestinian-on-Palestinian violence is increasing.

Jordan under King Hussein, and since 1999 under his son King Abdullah, has had to steer an often treacherous path between its Arab neighbors and its Western allies. Although King Hussein fiercely opposed the 1979 peace treaty between Egypt and Israel, he himself made peace with Israel in 1994. (Israel remains officially at war with Syria and Lebanon.) In part because of its dependence on Iraqi oil, Jordan sat out the first Gulf War in 1991, jeopardizing its relationship with the United States. The collapse of the Iraqi oil industry after the U.S-led invasion in 2003 has deprived Jordan of its main fuel source and put a heavy strain on its economy. Tourism has become an important sector of Jordan's economy, focusing on the spectacular ancient site of Petra and the stark natural beauty of Wadi Rum.

Lebanon and Syria

Since 1991, Lebanon has been putting itself together after a horrendous 16-year civil war, during which its capital, Beirut, once known as the Paris of the Middle East, was reduced more or less to rubble. Israel withdrew its troops from Southern Lebanon in 2000, but the Syrian- and Iranian-backed Shi'a group Hezbollah built a sizable force in South Lebanon. Although Lebanon was able to rebuild and was beginning to enjoy a new level of prosperity and hopes for peace, Hezbollah triggered an Israeli counterattack in July 2006 which unleashed a horrible cycle of death and destruction. The crisis is still unresolved, although Israeli forces left Lebanon under U.N. mandate and assurances of a U.N. peacekeeping force and the promise of Hezbollah giving up weapons. Unfortunately, Hezbollah has subsequently refused to disarm, and the situation is very tense, with Hezbollah conflicting with the elected government of Lebanon and threatening conflict with the Israelis.

The Ba'ath Party governs Syria. Since 1970 the country has been run dictatorially by Hafez al-Assad and after his death in 2000 by his son, Bashar. The long-standing conflict between Syria and Israel centers on Israel's capture of the Golan Heights from Syria in the 1967 Arab-Israeli War. Since the fall of Saddam Hussein in 2003, Syria has been under increasing pressure from the United States. Syria's economy is still largely state-controlled and although the country is an oil exporter, its economy has been struggling with high unemployment rates, a fast-growing population, and a variety of internal and external crises.

Iraq

From 1958, when its king was assassinated in a military coup, until April 9, 2003, when the government of Saddam Hussein was overthrown by a U.S. and British invasion, Iraq had been a military dictatorship ruled by the Arab socialist Ba'ath party. From 1979 on, Iraq was under the thumb of the cruel and repressive regime of Saddam Hussein. In 1980, Iraq launched an eight-year war against Iran that left some 1.5 million people dead. In 1990, Saddam Hussein invaded Kuwait and held it until January 1991, when his forces were routed by a U.S.–led United Nations coalition in under a week. Throughout the 1990s, U.N.–imposed sanctions curtailing Iraq's oil exports ruined the country's economy and caused great suffering to the Iraqi people. Yet Hussein would not fully cooperate with U.N. weapons inspectors seeking to verify the presence or absence of weapons of mass destruction in Iraq. Finally, in March 2003, U.S. president George W. Bush, incorrectly alleging that Iraq did have weapons of mass destruction and was prepared to use them, launched an invasion of Iraq that succeeded in overthrowing the Hussein regime.

Saddam Hussein's costly wars and the U.N. sanctions that followed the first Gulf War wrecked Iraq's oil-rich economy. (Iraq has the fourth largest oil reserves in the world.) Years of neglect under Saddam Hussein coupled with widespread post-war looting has seriously damaged much of Iraq's infrastructure. The U.S. occupation is proving considerably more difficult and costly than the Bush administration had estimated. Even after the capture of Saddam Hussein on December 13, 2003, insurgent attacks on U.S. and coalition forces have continued to increase. It remains to be seen whether Iraq will succeed in transforming itself into an open democratic society or become an even more dangerous threat to regional peace.

Ottoman and Persian Domains

The subregion we're calling the "outer" Middle East lies north and east of the Middle East Core and consists of just three countries: Turkey (capital: Ankara), Iran (Teheran), and the part Turkish/part Greek island of Cyprus (Nicosia). The two big countries, Turkey and Iran, share a number of traits. Both are predominantly Muslim countries, but neither is an Arab country. (Cyprus is neither Arab nor predominantly Muslim.) Both touch other regions. Iran borders South Asia (Pakistan), Central Asia (Afghanistan and Turkmenistan), and the Southern Caucasus countries. Turkey also borders the Caucasus and it not only borders Europe, it is part of it. Most of Turkey consists of the great peninsula called Asia Minor or Anatolia, which is definitely in Asia. But a small, triangular part of Turkey lies in Europe, across the Bosporus Strait, the Sea of Marmara, and the Dardanelles Strait from Asian Turkey. European Turkey is bordered on the west by Greece and Bulgaria.

The nations of this subregion border many of the region's major bodies of water. Iran touches the Caspian Sea on the north, and the Persian Gulf, the Strait of Hormuz, and the Gulf of Aden, on the west and south. Turkey is washed on the north by the Black Sea, on the west by the Aegean, and on the south by the Mediterranean, which surrounds the island of Cyprus.

Both Turkey and Iran have long and glorious histories that have often intersected. The Turkish Empire conquered the Byzantine capital of Constantinople in 1453 and renamed it Istanbul. The Turks ruled a far-reaching empire, but gradually declined and lost most of their territories. A republic was declared in 1918 and the capital moved to Ankara.

The Persian Empire had a long and glorious past. In 1979 the last monarch, Shah Reza Pahlevi was overthrown by the Islamic Revolution of Ayatollah Khomeini, and the nation continues under strict theocratic rule.

The Rugged and Populous Subregion

All three countries in the subregion are mountainous. The western part of Asian Turkey consists mainly of the semiarid Anatolian Plateau, rimmed by mountains on the north and south. Eastern Turkey is the most mountainous part of the country. The country's tallest peak (17,000 feet) is Mount Ararat, famous as the place where Noah's Ark came to rest. It rises near the eastern border with Armenia in a part of Turkey that is historically Armenian. Turkey's lowland areas are mainly narrow strips along its coasts. The island of Cyprus has two east-west mountain chains with a wide plain in between. Iran also consists of a large plateau, in this case on the east spilling over into Afghanistan, with low mountains on the Iranian side of the border and tall, volcanic mountains in the north and west.

Most of Turkey is temperate, with hot, dry summers and cold, rainy winters. More extreme weather prevails in the interior. The northeastern and western parts of the country get the most rain. The climate in Iran's mountainous area is subtropical, with temperatures and rainfall depending on elevation. Elsewhere the climate is continental, with hot summers and cold rainy winters. The central plateau and southeastern coast plain are arid.

The Mediterranean island nation of Cyprus, which enjoys hot, dry summers and cool, damp winters is unique in this mostly Muslim region. Most of the Cypriots are of Greek Christian origin (78 percent), but a minority are of Turkish Muslim heritage (18 percent), smaller groups of other types of Christians making up the other 4 percent. The population of the entire island is estimated to be around 835,000.

Turkey and Iran are the population giants of the Middle East and Central Asia region, with a combined population larger by far than any of the other subregions. There are 72.6 million Turks and 70 million Iranians. Turkey's capital, Ankara, has a population of over 4 million, but its largest city, Istanbul, is really huge, with 10 million people in its metropolitan area. Iran's capital is even bigger—Teheran's metro area with more than 12 million inhabitants make it the largest city in the entire region.

Turkey was once the center of a vast, multicultural empire, but it is a surprisingly homogeneous country today. Its population is 70 percent Turkish, 20 percent Kurdish, and 10 percent other mixed groups. Turkish is the official language; Kurdish, Arabic, Armenian, and Greek are also spoken. The ethnic and linguistic breakdown in Cyprus mirrors the political division of the island. About 77 percent of Cypriots are Greeks, who live in the southern part of the island; about 18 percent are Turks, who live in the northern part.

Unlike Turkey, Iran is an ethnic and linguistic potpourri, reflecting its history as an international crossroads. Slightly more than half of Iranians are ethnic Persians; nearly a quarter are Azerbaijanis. Smaller minority groups include Kurds, Arabs, Lurs, Balochs, and Turkmen. About 58 percent of Iranians speak Farsi (Persian) or a Farsi dialect. Also spoken are Turkic dialects, Turkish, Kurdish, Armenian, Balochi, Arabic, and Luri.

Almost everyone in Turkey and Iran is Muslim, but there are divisions within Islam and they are reflected here as well. Turkey is almost 100 percent Sunni Muslim, with tiny Christian and Jewish minorities. Iran, the Shi'a heartland, is almost 95 percent Shi'a Muslim and only 4 percent Sunni. A tiny fraction of the population is Zoroastrian, Christian, Jewish, or Baha'i.

Turkey

Between 1923 and his death in 1938, Kemal Atatürk, the father of the Turkish republic, laid the foundations for a modern, secular Turkish state. Since 1938, Turkey has gradually developed into a parliamentary democracy despite periods of military rule and a lingering tendency on the part of the military to intervene in political affairs. Turkey has a troubled relationship with its own large and oppressed Kurdish minority. In 1984, a Kurdish separatist group launched a guerilla/terrorist campaign in southeast Turkey. Although a ceasefire was declared in 1999, skirmishes continue between government forces and Kurdish militants based in Iraq. The prospect of an independent Kurdistan in post-Saddam Northern Iraq is a source of great concern to the Turkish government given the size and discontent of its own Kurdish minority.

Geographically Speaking

The Kurds are a nation without a country. A non-Arab, mainly Sunni Muslim people, the Kurds inhabit a region that is unofficially called Kurdistan, encompassing Southeastern Turkey, Northern Iraq, Northwestern Iran, and small parts of Northeastern Syria and Armenia. Turkey has long oppressed its large Kurdish minority, although some of its harsher measures are being repealed as part of Turkey's bid to join the European Union. Saddam Hussein attacked Iraqi Kurdish villages with poison gas during the Iran-Iraq War in the 1980s and brutally put down a Kurdish uprising following the Persian Gulf War of 1991. A U.S.-enforced no-fly zone over Northern Iraq enabled the Kurds to set up an autonomous region, but fighting soon broke between Kurdish factions over control of the area. A cease-fire was declared in 1999. Kurdish forces assisted U.S. and British forces in the seizure of Kurdish villages during the U.S.-led invasion of Iraq in 2003.

Turkey's economy is a mix of old and new. A modern, dynamic industrial and commercial sector coexists with a traditional agriculture sector, which accounts for nearly 40 percent of the Turkish workforce. Turkey has oil and gas reserves, but not on the level of Iran and the other Middle Eastern oil giants. Turkey continues to press its application to join the European Union (though membership talks were partially frozen by the EU in December 2006).

Cyprus

Strategically located Cyprus has passed through many hands in its long history. Since its independence in 1960 tensions between Greek and Turkish Cypriotes continued to rise, with the Greek side pushing for union with Greece and the Turkish side aiming for partition. With civil strife and violence mounting, Turkey invaded Cyprus in 1974 and took over more than 30 percent of the island, displacing some 200,000 Greeks. Cyprus remains a divided island with an internationally recognized government on the Greek side and a Turkish Republic of Northern Cyprus, recognized only by Turkey, on the northern end. The Greek part of the island became a member of the European Union in 2004. The Greek Cypriot side is far more prosperous than the Turkish side, which is hampered by its lack of international recognition. The Turkish side is heavily dependent on agriculture, whereas the Greek Cyprus has a more diversified economy oriented to industry and the service sector. Scarcity of drinking water is a problem on both sides.

Iran

After 25 years of restrictive religious rule in Iran, a younger generation is increasingly pressing for change. Yet U.S.-Iranian relations remain troubled. In early 2003, President George W. Bush included Iran in his famous "axis of evil" along with Iraq and North Korea. After the Iraq invasion and the overthrow of Saddam Hussein, Iran is facing intense U.S. and international pressure to cooperate with U.N. weapons inspectors and shut down the nuclear weapons development program it is alleged to have.

The brutal Iran-Iraq War of the 1980s, in which both sides used chemical weapons, had a devastating effect on the economies and oil industries of both countries. Unlike Iraq, however, Iran has repaired much of the damage. Iran has abundant natural resources, but oil is king (second largest reserves of conventional crude oil in the world and second in the world in natural gas reserves), accounting for 85 percent of the country's exports. Other major exports include petrochemicals, textiles, Iran's famous Persian carpets, fruits and nuts, and iron and steel. Inflation, unemployment, and poverty rates all remain high.

At the Edge of Asia: Transcaucasia

The Caucasus Mountains are an east-west chain that marks the dividing line between Asia and Europe. The greater Caucasus region straddles the mountains and is divided into two parts: North Caucasia, which remained part of Russia after the split-up of the Soviet Union in 1991 (see Chapter 16) and Transcaucasia, the southern (actually southwestern) part of the region covered here. The three nations that make up this subregion—Armenia (its capital is Yerevan), Azerbaijan (Baku), and Georgia (Tbilisi)— are all former Soviet republics that spun off and formed independent nations in 1991.

Transcaucasia is dominated by the southern slopes of the Caucasus Mountains and the Azerbaijan Plateau in the east. The glacier-fed Kur River flows through the plateau into the Caspian Sea. Famous for its scenic beauty, the Caucasus region is prone to occasional earthquakes.

Georgia's climate along its Black Sea coast is subtropical with heavy rains. It's cooler and dryer in the eastern plains, with more extreme conditions in the eastern mountains. Bordering on the Caspian Sea, Azerbaijan has a dry subtropical climate with hot summers and mild winters. Totally landlocked Armenia has a highland continental climate with hot summers and cold winters.

Azerbaijan is the most populous of the three Transcaucasian countries with 8.4 million people. Georgia is second with 4.7 million, and Armenia third with 3.2 million. Azerbaijanis are about 90 percent ethnic Azeri, with small minorities of Dagestanis, Russians, and Armenians. Azerbaijani is the predominant language, followed in a small way by Russian and Armenian. About 94 percent of Azerbaijanis are nominally Muslim; small percentages are Russian or Armenian Orthodox. The Georgian population is about 70 percent ethnic Georgian; Armenians, Russians, Azeris, Ossetians, and Abkhas are also present in smaller numbers. The predominant religion is Georgian Orthodox, with sizeable Muslim, Russian Orthodox, and Armenian Apostolic minorities. The people of Armenia are about 95 percent Armenian in ethnicity, language, and most practice the Armenian Apostolic Church religion.

The Caucasus have been invaded and settled by many different races and cultures over the centuries. Persians, Khazars, Arabs, Huns, Turko-Mongols, Ottomans, and Russians have all left their imprint on the Caucasus. The three Transcaucasian countries haven't had an easy time of it since independence. Azerbaijan has lots of oil and even greater untapped reserves, but is held back by Soviet-style inefficiency, corruption, and slow progress in opening up its economy. Another major problem is the ongoing struggle with Armenia over Nagorno-Karabakh, a small bit of Azerbaijani territory with a largely Armenian population, which Armenia seized in 1994, along with a chunk of Azerbaijan proper. Armenia's economy has also been badly affected by the conflict, especially as it has no oil and scant mineral deposits, and has to import food. Armenia, like the rest of this region, is struggling to de-Sovietize and jumpstart its energy- and resource-poor economy. Georgia has also had to cope with ethnic and civil strife in two Muslim breakaway areas: South Ossetia and Abkhazia.

Central Asia: Afghanistan and the -stans

Afghanistan is the former Taliban stronghold and al-Qaeda terrorist training camp. The other five countries in the region are the former Soviet republics of Central Asia: Kazakhstan (capital: Astana), Kyrgyzstan (Bishkek), Tajikistan (Dushanbe), Turkmenistan (Ashkhabad), and Uzbekistan (Tashkent).

One thing all these countries share is the syllable -stan or -istan at the end of their names. *Stan* is an Old Persian word for "place of," usually preceded by the name of a particular tribe (Kazakhstan, for example, is the place or homeland of the Kazakhs), which gives you an indication of the tribal nature of the societies in this area. The huge subregion of Central Asia abuts Russia to the north and China to the east. Pakistan and Iran are on the subregion's southern border; the Caspian Sea and Russia lie to the west. Although Turkmenistan and Kazakhstan border the Caspian Sea, all the Central Asian countries are landlocked.

Mountains and Deserts

Beginning in the north, giant Kazakhstan (it's about four times the size of Texas) is mostly one vast, flat expanse of arid grasslands with high mountains rising in the southeast. The high, glacier-clad Tian Shan and Pamir Mountain ranges dominate Kyrgyzstan and Tajikistan. (At 24,590 feet, Pik Imeni Ismail Sameni in the Pamir range in Tajikistan is one of the world's highest mountains.) The Kyzyl Kum desert (*kum* is a Turkish word meaning "sand") covers most of central Uzbekistan; to the east and southeast are fertile valleys. Almost 90 percent of Turkmenistan is covered by the Kara Kum desert. From its center to its eastern border with Pakistan, Afghanistan is a very mountainous country. Its tallest peaks, some rising above 24,000 feet, rise up in the northeastern Hindu Kush range, home of the famed Khyber Pass (see Chapter 23). Southern and Southwestern Afghanistan is mostly desert.

This immense area has a relatively small population. With almost 30 million people, Afghanistan is Central Asia's most populated country. Uzbekistan comes next with 26.5 million, followed by Kazakhstan with almost more than 15 million and Tajikistan with 6.7 million. Kyrgyzstan and Turkmenistan each have a bit around 5 million.

Central Asia—physical features.

The Ethnic Web

The names of the countries in this subregion let you know who the predominant ethnic group is. In the five former Soviet republics, you're also likely to find a Russian minority of varying size. In addition, the population of each of these countries includes people from most of the other countries in the subregion as well as people from outside it. The majority Muslim populations of all of these countries are overwhelmingly Sunni. But it's not that simple. The population of Kazakhstan is about 50 percent Sunni Muslim Kazakh and 30 percent Russian (mostly Russian Orthodox), with smaller minorities of Ukrainians, Germans, Uzbeks, and Tatars. Uzbekistan is about 88 percent Sunni Muslim Uzbek and about 9 percent Russian Orthodox Russian, with tiny Tajik, Kazakh, and Tatar minorities. In Afghanistan the largest group is the

Pashtuns (ethnic Afghans), who live in the eastern and south central parts of the country. Baluchis inhabit the deep south. Tajiks live near the Tajikistan border, Uzbeks near the Uzbekistan border, and Turkmen near the Turkmenistan border. Hazaras, of Mongolian origin, live in the mountainous center. All these groups are mostly Sunni Muslim, except for the Hazaras, who are Shi'a.

Afghanistan

Afghanistan was neutral in World Wars I and II and throughout the cold war until the late 1970s. In 1979 the Soviet Union invaded Afghanistan to support the country's new and increasingly unpopular Marxist regime. The 10-year war that followed pitted Soviet forces and Soviet-backed Afghan government troops against U.S.–backed Afghan *mujahedeen* (Arabic for "holy warriors"). The war took a horrific toll on both sides, but the mujahedeen finally prevailed and the Soviets withdrew in 1989. The pro-Soviet government fell in 1992, and the chaos that followed gave an opening to a group of radical Islamic students called the Taliban. After taking control of Kabul in September 1996, the Taliban began imposing on Afghanistan a fundamentalist Islamic regime of unprecedented harshness. Under the Taliban, the country became a haven for the Islamist terrorist group al-Qaeda and its leader, Osama bin Laden.

After the September 11, 2001, al-Qaeda attacks on the World Trade Center in New York and the Pentagon in Washington, D.C., the Taliban refused to hand over bin Laden. On October 7, the United States and its allies launched an air war against the Taliban regime. Within five weeks, anti-Taliban Afghan forces on the ground had taken the key cities of Mazar-i-Sharif and the capital, Kabul. The Taliban government fell on December 7, but its leader, Mullah Omar and, more important, Osama bin Laden were not caught, and remain at large. Today, a major problem is the ongoing production of opium. Poor Afghanis can make a fortune growing opium, whereas their farms can barely sustain them if they grow foodstuffs on their land. As a result, many turn to the illegal drug trade, which accounts for one third of the Afghan GDP, and makes Afghanistan the number-one opium producer in the world. It is believed that much of this illegal trade directly sponsors terrorism and the Taliban insurgency.

The Five -stans

Like the former Soviet republics of Eastern Europe, the Central Asian republics are struggling with varying degrees of success to reform and open their political systems and economies. Kazakhstan, the second largest of the former Soviet republics after

Russia, is rich in fossil fuels and other natural resources and has a healthy agricultural sector and a growing industrial sector. Its per capita GDP is the strongest in the region.

Turkmenistan also has major oil reserves and even more significant natural gas reserves (the fifth largest in the world). And even though it's a mostly desert country, Turkmenistan is the world's tenth largest cotton producer, thanks to intensive irrigation. Its economy, however, is seriously hampered by a repressive ex-communist government that doesn't tolerate opposition and has been reluctant to introduce much-needed market reforms.

Kyrgyzstan was the first of the Central Asian republics to adopt democratic institutions and the first of the ex-Soviet republics to join the World Trade Organization in 1998. It is a poor, predominantly agricultural country, with significant gold, mercury, uranium, and natural gas deposits. Election fraud, a high poverty rate, ethnic strife, and Islamic insurgencies are among the problems Kyrgyzstan is grappling with.

Uzbekistan is the world's second largest cotton producer, but that comes at a cost. Intensive irrigation and overuse of chemical fertilizers have poisoned an already arid land and dried up water supplies like the Aral Sea. Uzbekistan is also an exporter of gold and has other significant mineral and fossil fuel reserves, but its future is clouded by problems similar to those of Kyrgyzstan.

Despite healthy growth in recent years, Tajikistan is still the poorest of all the former Soviet republics. A civil war that lasted through most of the 1990s wreaked havoc with an already fragile, mainly agricultural economy. As elsewhere in the region, cotton is king in Tajikistan.

The Least You Need to Know

- The Middle East is at the crossroads of Europe, Asia, and Africa.
- The Middle East Core and the Arabian peninsula gave rise to three of the world's major religions: Judaism, Christianity, and Islam.
- In Central Asia, the former Soviet republics are struggling to liberalize their economies and political systems.
- The entire Middle East and Central Asia region has become a battleground in the global war on terror.

Chapter 23

South Asia: The Mighty Subcontinent

In This Chapter

- ◆ Looking back at South Asia's history
- ◆ Measuring the mountains and other natural features
- ◆ Discovering India, Pakistan, and Bangladesh
- ◆ Exploring Nepal and Bhutan
- ◆ Cruising through the island countries

The region of South Asia, as shown on the following map, includes the countries of Pakistan, India, Nepal, Bhutan, Bangladesh, Sri Lanka, and the Maldives, in other words, the Indian subcontinent and its immediate neighbors.

South Asia is walled off on the north by the world's highest and mightiest mountain chain, the Himalayas and the country of China. To the east is Myanmar (formerly Burma) and the Bay of Bengal; to the south, the Indian Ocean; and to the west, the Arabian Sea, Iran, and Afghanistan. After the Indian plate broke off from Eastern Africa and Antarctica, it began drifting northeast until it crashed into the Eurasian plate. This process caused the Himalayan upsurge and continues to make the Himalayas rise.

South Asia is small in area compared to some other regions, but it has a huge population. With more than 1.5 billion people, South Asia is running neck and neck with East Asia (China, Mongolia, the Koreas, and Taiwan) for the title of "most populous region in the world"; and with its higher rates of population growth, it is destined to win the race.

The Very Long Road to the Present

One of the world's four Bronze Age urban cultures (along with Egypt, Mesopotamia, and China), the Indus Valley civilization flourished in the fertile Indus River basin in the western part of South Asia (today's Pakistan and Western India) from about 2600 B.C.E. to about 1600 B.C.E. Between 1600 and 1500 B.C.E. Aryan tribes from the northwest overcame what was left of the Indus civilization and settled in the Indus Valley region. The Aryans introduced the Sanskrit language, the Vedic religion, a precursor of Hinduism, and the caste system.

Geographically Speaking

India's caste system is a hereditary social structure closely bound up with Hindu beliefs about personal conduct and reincarnation. The four main castes (each subdivided into many subcastes) are, in descending order: Brahmins (priests and religious officials), Kshatriyas (warriors, rulers, and large landowners), Vaisyas (farmers, merchants, traders, and artisans), and Shudras (servants and peasants). At the bottom of the social order, beyond the caste system (or "outcastes," which is how the word originated) were the Untouchables or "harrijan," fit only for the most menial of tasks.

Under the caste system, people are born into the caste of their parents and are restricted to certain occupations. Marriage and intermingling across castes are forbidden. India's constitution bans caste-based discrimination, but the caste system, and its injustices, persist, especially in the countryside and rural villages that are the backbone of Indian life.

Buddhism was founded in India in the sixth century B.C.E. by Gautama Siddharta (the "Buddha"), who was born in what is now Nepal in 563 B.C.E. The new religion spread throughout Northern India, especially during the reign of Ashoka, the great king of the Maurya dynasty, who unified most of India for the first time in the third century B.C.E.

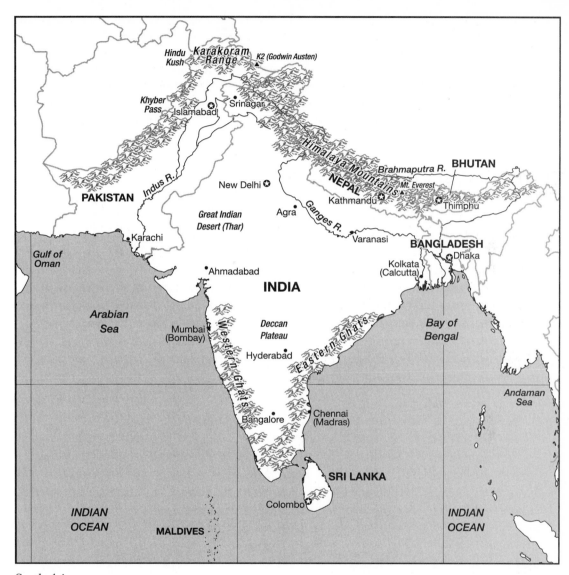

South Asia.

The 33-mile-long Khyber Pass (elevation 3,500 feet) provides a passage through the rugged Hindu Kush Mountains that separate Afghanistan and Pakistan. Over the centuries, many conquerors and would-be conquerors have used the Khyber Pass to invade India, including Alexander the Great, Huns, Arab Moslems, and, eventually, various Europeans starting in 1499. India became a part of the British Empire in 1858.

> **Terra-Trivia**
>
> In 1633, the Mughal emperor Shah Jahan commissioned one of the world's most beautiful buildings and certainly the most famous structure in India today, the Taj Mahal, in Agra, near Delhi. A love letter written in gleaming white marble, the Taj Mahal was designed as a mausoleum for the emperor's beloved wife, Mumtaz Mahal, who had died giving birth to their fourteenth child.

Independence and Partition

The British were not kind masters of the subcontinent. They put down rebellion with an iron fist, but they could not combat the freedom movement initiated by Mohandas Gandhi, the "Mahatma," or "Great Soul." British India was granted a measure of self-rule in 1935; it attained full independence in 1947.

Independence, however, was accompanied by a violent partition of British India into two nations: India, with a majority Hindu population, and the smaller Muslim state of Pakistan, which consisted of two separate parts on either side of Northern India. In one of the largest migrations in history, some 17 million people fled across the Indo-Pakistani borders in both directions to avoid sectarian rioting and bloodshed that was taking place on both sides. Millions died in the process.

A still unresolved conflict over the status of the northwestern Indian state of Jammu and Kashmir led to warfare between the two new countries in 1947, in 1965, and in 1999. In 1971, India supported an independence movement in East Pakistan, which, after yet another Indo-Pakistani war, became the independent nation of Bangladesh. Hostility between India and Pakistan and between Muslims and Hindus in both countries continue to make the subcontinent a dangerous tinderbox, especially since 1998, when first India and then Pakistan tested nuclear weapons.

From the Bay of Bengal to the Top of the World

South Asia lines up latitudinally with Saharan Africa and Middle America but has little in common with either. A region of great contrasts, South Asia has virtually every physical feature except major lakes. Below the Himalayas and their foothills, the Asian subcontinent encompasses the fertile Indus Valley and the Great Indian (or Thar) Desert in the west, the Ganges basin in the center, and the Brahmaputra River in the east.

The great Deccan Plateau dominates the center of the subcontinent peninsula. The Deccan is flanked by the Western and Eastern Ghats (steep slopes) down to the narrow coastal plains. Rugged and treacherous mountains separate the Indus from Afghanistan, and rain forest hills separate the Brahmaputra from Myanmar.

Although temperature varies with the seasons, it's colder in the mountains and plateaus. The Himalaya Mountains average only 12 degrees F, whereas central India averages 79 degrees F. Rainfall also varies with topography. North of Bangladesh, Cherrapunji gets 450 inches per year (second only to Mount Waialeale in Kauai, Hawaii, with 460 inches). The Great Indian Desert receives only 4 inches; the rain forest in the east, 80 inches; the plain, 40 to 80 inches; and almost everywhere else, 20 to 40 inches.

The most famous of South Asia's many large rivers are the Indus (1,800 miles long), the main river of Pakistan; the Hindu holy river, the Ganges (almost 1,600 miles long), flowing from north central India into Bangladesh; and the Brahmaputra (almost 1,800 miles long), in Northeastern India and Bangladesh. All three rivers originate in the mountains of Tibet to the north and end in great deltas. The Indus River delta is on the Arabian Sea, and the Ganges and Brahmaputra join in Bangladesh to form a vast delta on the Bay of Bengal.

South Asia's Titan: India

India is a large country with a land area about one third that of the United States. Its population of more than a billion is almost four times the population of the United States. India today is home to about one sixth of the world's population.

India has many cities with more than a million inhabitants. The largest metropolitan areas are Delhi with an estimated population of 15 million, and the financial capital, Mumbai (Bombay), whose population is about 13 million. India's capital city is New Delhi, which makes up a small part (population about 320,000) within the metropolitan Delhi conurbation. India's ethnic groups have so intermingled over the centuries that determining who's what is difficult. A widely accepted generalization is that about 72 percent of the population

Terra-Trivia

In recent years, many Indian cities, including three of its megacities, officially replaced their colonial names with the local language version of their names. Bombay became Mumbai in 1995, Madras was changed to Chennai in 1996, and in 2000 Calcutta became Kolkata.

is Indo-Aryan, 25 percent is Dravidian, and 3 percent is "other." Indo-Aryans are the result of the ancient intermingling between local populations and the Aryan tribes that migrated to India from the north around 1500 B.C.E. Indo-Aryans predominate in the north where Aryan influence was strongest. Darker skinned Dravidians, descendents of pre-Aryan Indian populations, predominate in the south.

Hindi is India's main official language, but it's the native tongue of only about a third of the population. In addition to Hindi and classical Sanskrit, the Indian government recognizes 16 other regional languages as official. English enjoys "associate" status in the language hierarchy, but it is spoken by most educated Indians and is used for many national, political, and commercial communications. India's scores of languages and countless dialects fall into four main language families, with the majority of Indians speaking either an Indo-Aryan or a Dravidian tongue. (The linguistic north/ south divide is much sharper in India than the racial divide.) The most widely spoken Indo-Aryan languages after Hindi are Bengali, Marathi, Urdu, Gujarati, Oriya, and Punjabi. Telugu, Tamil, Kannada, and Malayalam are the main Dravidian languages of the south. Many of India's languages are mutually unintelligible, which is why Hindi and English are so important.

Just over 80 percent of the Indian population is Hindu; about 12 percent is Muslim. Christians of all denominations account for 2.3 percent of the population. Less than 2 percent of Indians are Sikhs, practitioners of a monotheistic religion founded in Punjab in the fifteenth century C.E. Although both originated in India, Buddhism and Jainism (a reform movement in Hinduism that became a separate religion in the sixth century B.C.E.) today account for small fractions of the Indian population. Despite their small numbers, however, Jains play a disproportionately large role in the economic, political, and cultural life of India.

India After Independence

The miracle of modern India is that despite its many problems—massive overpopulation; widespread poverty; a hostile and nuclear-armed neighbor; and continuing ethnic, religious, and caste tensions within its borders—it has preserved the rule of law and its democratic institutions for more than half a century.

India Today

Behind the common perception of India as a place of grinding, if colorful, poverty lies a far more complex economic reality. By any definition, India is a very poor country.

About a quarter of the population lives below the poverty line, and for India's poor, even finding adequate nutrition is often a daily struggle. Agriculture looms very large in India's economy, contributing about 25 percent of GDP and employing a whopping 60 percent of its workers, most of them traditional village farmers. India also has the world's largest middle class, as many as 200 million people with good jobs and disposable income. One of India's most important resources is its large pool of well-educated, English-speaking workers.

India's main industries include textile manufacturing, chemicals, food processing, steel and heavy equipment manufacturing, coal and iron mining, and petroleum and natural gas production. In recent years, it has become a major exporter of software services and workers. The country has a lively press and publishing industry and the world's largest motion-picture industry (yes, bigger than Hollywood in terms of the number of films produced per year!).

Despite impressive economic gains in recent years, however, India's economy is still weighed down by a poor infrastructure, inequitable income distribution, a cumbersome bureaucracy, and the enduring problems of overpopulation and Hindu-Muslim sectarian violence.

Terra-Trivia

One of the great legacies of British rule in India is its remarkable railroad system. With nearly 39,000 miles of track, it is the largest railroad system in Asia, the second-largest state-owned system under a single management in the world, and the fourth most heavily used.

Pakistan: Born and Bred in Conflict

The easternmost of the -stans, Pakistan is bounded by India on the east, China and Afghanistan on the north, Afghanistan and Iran on the west, and the Arabian Sea on the south. Northern and Western Pakistan are very mountainous; the Karakoram Range along the Chinese border and Hindu Kush along the Afghan border contain some of the highest mountains on Earth. In the southwest are the Baluchistan highlands, a forbidding landscape of deserts, rocky plains, and mountains. The eastern half of Pakistan is flat and dominated by the Indus River and its tributaries. This alluvial plain is shared by two provinces, the Sind in the south and the Punjab, the Pakistani heartland, in the north. East of the plain, straddling the Indo-Pakistani border is the Great Indian, or Thar, Desert. Southern Pakistan is very dry; the north, although far from lush, receives slightly more precipitation.

Pakistan's 169 million people make it the sixth most populous country in the world, and the second most populous Muslim country, after Indonesia. (India's Muslim minority is almost as large as Pakistan's entire population.) The vast majority of Pakistanis, about 77 percent, are Sunni Muslims; about 20 percent are Shi'a. Urdu is the official national language, although it is the native tongue of only about 8 percent of Pakistanis. The language of Punjab province, Punjabi, is spoken by nearly half of the population. The other widely spoken languages are Sindhi, Siraiki, Pashto, and Baluchi. English is also an official language and used by government ministries and the educated elite.

Pakistan's capital is Islamabad (about 950,000 people). It was built in the 1960s to replace the old capital of Karachi (population 14.5 million), which is Pakistan's largest city and main seaport and industrial center.

Post-Partition Pakistan

Pakistan has had a very troubled history since independence, which includes corruption, military coups, elections, and more coups.

Despite the entrepreneurial skills of its people and its considerable natural resources—fertile land, natural gas, coal, some petroleum, and large hydroelectric potential—Pakistan remains poor and underdeveloped, its economy hampered by financial mismanagement, political instability, corruption, a low literacy rate (46 percent), and the costly ongoing conflict with India. International sanctions imposed after Pakistan conducted nuclear bomb tests in 1998 were a further blow to its economy. Agriculture accounts for 24 percent of Pakistan's GDP and employs about half its workforce. Textile and garment manufacturing is its major industry.

The support of Pakistan's leader, General Pervez Musharraf, was crucial to the war on terrorism declared by U.S. president George W. Bush after the September 11, 2001, al-Qaeda attacks in the United States. To shore up that support, the nuclear test-related sanctions were lifted and Pakistan received a massive infusion of foreign aid.

The Kashmir Conflict

Before partition, Kashmir was one of India's so-called princely states, nominally ruled by a Maharajah. In this case, the Maharajah was Hindu, but most of the population was Muslim. When it came time to separate Pakistan from India, Pakistan invaded Kashmir and the Maharajah turned to India for help. Indian forces counterattacked and at the end of the war in 1949 Kashmir was divided between Pakistan and India, with India retaining the capital, Srinigar. In 1956, despite Pakistani protests, India-controlled Kashmir became the Indian state of Jammu and Kashmir, the country's only Muslim-majority state. The continuing dispute over Kashmir led India and Pakistan to war in 1965 and in 1999 and to the brink of war many more times, most recently in 2001, after a suicide bombing in the Indian Parliament killed 14 people, and in 2003, after 24 Hindus were killed in Indian-controlled Kashmir.

Disaster-Prone Bangladesh

The cultural, economic, political, not to mention physical divide between East and West Pakistan eventually grew too great, and in 1971, East Pakistan declared its independence. After an extremely costly civil war (an estimated 1 million dead and 10 million refugees), East Pakistan prevailed and became the independent state of Bangladesh. Its modern history is also marred by coups and intermittent attempts at democracy.

Geographically Speaking

In India and Sri Lanka, as well as in Bangladesh and Pakistan, women are subjected to such atrocities as honor killings, dowry deaths (for insufficient dowries), disfiguring acid attacks, and female infanticide.

Ironically, all four countries have been pioneers in electing women to the highest offices in the land. Indira Gandhi was prime minister of India in the 1970s and '80s. Benazir Bhutto was Pakistan's prime minister in the 1980s and '90s. For most of the period from 1991 to the present (mid-2003), Bangladesh has been governed by two female prime ministers. In 1960, Sri Lanka became the first country in the world to have a female prime minister and since 1994 has had a female president.

Bangladesh is bordered on the south by the Bay of Bengal and by India everywhere else, except for a small stretch in the southeast where it borders Myanmar (Burma). Aside from a few hills and forests along its perimeter, Bangladesh is a low-lying country, consisting mostly of the double delta of the Ganges and Brahmaputra rivers. Its climate is tropical monsoon, with a hot and rainy summer and a dry winter. Bangladesh is one of the wettest places in the world, with most of its rainfall occurring in the monsoon season from June through September. Since independence, Bangladesh has experienced catastrophic floods requiring emergency foreign assistance in 1974, 1984, 1987, 1988, and 1998. In the 1998 flood, nearly 70 percent of the country was under water for several months and some 1,200 people died. Flooding in Bangladesh has been exacerbated by the severe forest cutting in the foothills of the Himalayas.

> **Eye on the Environment**
>
> As if Bangladesh did not already have enough water problems, naturally occurring arsenic is contaminating its groundwater supply, a primary source of drinking water. Inadequate well-water testing kits are compounding the looming public-health crisis.

Bangladesh is also one of the world's most cyclone-prone countries. Fierce tropical cyclones funnel up the Bay of Bengal to the coastal districts of Bangladesh at an average rate of 1.3 a year. In 1991, a cyclone killed almost 140,000 people and caused $2.7 billion in damages. Since then, however, Bangladesh has greatly improved its cyclone preparedness by building shelters and implementing evacuation plans.

Crammed into every inch of livable space are 147 million people, making Bangladesh the most crowded large country on Earth and the seventh most populous. Dhaka, the capital and largest city of Bangladesh, has more than 11 million people in its metropolitan area. The second largest city, Chittagong, has a population of about 6 million.

Bangladesh is an extremely poor and underdeveloped country. Bangladesh's per capita GDP of $2,100 is in the sub-Saharan African range. Barely 41 percent of the total population are literate, about 40 percent are unemployed or underemployed, and an enormous 60 percent are engaged in agriculture. On the positive side, Bangladesh has plenty of fertile and well-watered (to say the least) land and is now self-sufficient in rice. Its other main crop and a major export product is jute, used to make burlap, twine, and rope. Bangladesh also has substantial natural gas reserves but has been slow to develop them. Despite determined domestic and international efforts to boost its prospects, the hurdles for Bangladesh remain high.

Mountain Monarchies and Island Nations

In addition to the three nations that came out of the partition of British India, South Asia includes two Himalayan kingdoms (it used to be three, until the tiny Kingdom of Sikkim became India's smallest state in 1975) and two tropical island nations.

Nepal

This small landlocked country in the Himalayas is sandwiched between Indian and China's Tibetan Autonomous Region. Northern Nepal is dominated by the Himalayas: 8 of the 10 highest mountain peaks in the world are in this part of the country, including the world's highest, Mount Everest (29,028 feet), which straddles the Nepal-China border. The center of the country is occupied by the Himalayan foothills; the southern Terai region is a flat and fertile plain. The mountains are snowy and cold year-round and the foothills are cool. The Terai is hot and humid in the summer and cool in the winter.

Nepal's 57,000-square-mile area supports a population of nearly 27 million. Its fabled capital of Kathmandu, located in a broad valley in the central hills, has a metropolitan area population of 1.5 million. Nepal is home to many different caste and ethnic groups (the two terms are often used interchangeably here). The Sherpas, a Tibeto-Nepalese people living in the sparsely populated mountain north, have gained worldwide fame as expert Himalayan expedition guides.

Nepali is the official language of the country, spoken by about 90 percent of the population. About a dozen other languages and many more dialects are also spoken. Nepalese in business and government often speak English as well. Although Nepal was the birthplace of the Buddha, barely 8 percent of Nepal's people are Buddhist. The vast majority (more than 86 percent) is Hindu (though in Nepal there is a strong intermingling of Hindu and Buddhist beliefs). There are also small Muslim and Christian minorities.

Nepal successfully avoided being taken over by Great Britain during the colonial period, but at the cost of remaining economically backward. This mountain monarchy has a Maoist guerrilla movement that has been waging a growing and increasingly violent revolt against the Nepalese government since 1996. Peace talks in 2001 collapsed, but after another cease-fire in January 2003, negotiations resumed in April. However, the nation remains troubled and the king's government is becoming increasingly unpopular. Nepal's per capita GDP ($1,400) is even lower than that of Bangladesh. More than 80 percent of its people depend on agriculture for their livelihood, and about 42 percent live below the poverty line. In addition to farming, Nepal's main economic activities are tourism, textile manufacturing, and carpet production. The country's infant mortality rate is high and its life expectancy short (just 59 years).

Eye on the Environment

One of the most serious environmental problems facing Nepal, and much of South Asia, is deforestation. The consequences of unchecked deforestation are readily visible in Nepal and elsewhere in South Asia: severe soil erosion, water pollution, increased flooding and drought, and loss of wildlife. Nepal's forests shelter some of the world's rarest animals, including Bengal tigers, one-horned rhinos, and snow leopards, all of which are at risk.

Bhutan

East of Nepal, and separated from it by a small strip of India that was formerly the Kingdom of Sikkim, lies South Asia's other Himalayan kingdom, Bhutan. Like Nepal, Bhutan is bordered on the north and east by China (Tibet) and by India everywhere else. It is mostly mountainous, with some fertile valleys and a small stretch of plain in the south called the Duars. The climate is tropical in the southern plain and gets progressively colder the higher up you go. Measuring only 18,000 square miles (less than a third the size of Nepal), Bhutan is home to approximately 672,000 people. Its capital, Thimphu, has a population of 60,000.

Ethnically, the majority of Bhutanese are either Bhote (50 percent) or Nepalese (35 percent). The official language is Dzongkha, although a number of Tibetan and Nepalese dialects are also spoken. Seventy percent of the Bhutanese are Buddhist; most of the rest are Hindu.

Bhutan was largely isolated from the rest of the world until the 1960s, when it began opening up to the international community and launched a modernization campaign that included the abolition of slavery and the caste system. In 1998, Bhutan's king voluntarily transferred executive powers to an elected Council of Ministers. After nearly a century of absolute monarchy, Bhutan is slowly moving toward becoming a parliamentary democracy.

Agriculture and forestry are the mainstays of Bhutan's extremely small and under-developed economy. Most Bhutanese are subsistence farmers, and the country relies heavily on trade with and financial aid from India, which also handles its foreign affairs. Bhutan's resources include hydropower (it now exports electricity to India) and tourist appeal, which remains mostly unexploited.

Sri Lanka

Formerly known as Ceylon, Sri Lanka is the tear-shaped island off the southeast tip of India. The south-central interior of the island is mountainous; the rest is flat to rolling plains ending in beautiful palm-fringed beaches. Because Sri Lanka is only 6 degrees north of the equator, it has a tropical monsoon climate.

Sri Lanka is home to nearly 21 million people. About 2.5 million of them live in the metropolitan area of the capital city, Colombo. (Sri Lanka also has a legislative and judicial capital with a tongue twister of a name: Sri Jayewardenepura Kotte.) In recent years, Sri Lanka's economy has become much more focused on industry (textiles and garments) and services, and far less so on plantation crops (rubber and tea). Despite a disastrous ethnic civil war, Sri Lanka has experienced solid economic growth over the last decade and has South Asia's highest per capita GDP.

The defining fact of Sri Lankan history has been the division and enmity between its two major ethnic groups. Nearly 75 percent of Sri Lankans are Sinhalese (Buddhist) while one fifth are Tamil (Hindu). Sinhala is the official language, but Tamil is also a national language. (English is often used for government business.)

The island gained its independence from colonial Great Britain in 1948, and changed its name to Sri Lanka in 1972. In 1983, longstanding Tamil frustrations at the political and economic control of the majority Sinhalese reached a boiling point. A violent war broke out between government forces and Tamil rebel groups demanding a separate nation. The most notorious of the rebel groups is the Liberation Tigers of Tamil Eelam, or Tamil Tigers for short, pioneers in the use of suicide bombings. Finally, in December 2001, after tens of thousands of Sri Lankans had lost their lives (62,000 by some estimates), a ceasefire was declared. Trouble continues between these two groups.

The Maldives

The Maldives are an island republic in the north Indian Ocean, about 400 miles southwest of Sri Lanka. The country consists of about 1,190 coral islands on 26 atolls. (An atoll is a large ring of coral atop a defunct volcano surrounding a lagoon.) Although the total area of the islands is only about 115 square miles, they are spread out over 35,200 square miles. Only about 200 of the islands are inhabited by the Maldives' 329,000 people. (Another 87 "uninhabited" islands are set aside for tourist resorts.) Maldivians are of Sinhalese, Dravidian (South Indian), Arab, and African origin. All of them speak Divehi (a Sinhalese derivative), and all are Sunni Muslim. The capital, Male (that's its name, not its gender), has more than 80,000 residents.

The Maldives have superb beaches, magnificent coral reefs, and a tropical monsoon climate; tourism is the country's main industry. With little arable land, the Maldives must import most of its food. In a country whose highest elevation is barely six feet, global warming and rising sea levels are a major concern.

In the News: Pakistan's Northwest Frontier Province

Kashmir is not Pakistan's only international trouble spot. In many respects, its Northwest Frontier Province is an even more volatile and dangerous place. A wild and rugged area along the Afghan border, the Northwest Frontier Province is controlled in reality by local tribes and only nominally by the central government. As of mid-2003, it was widely believed that al-Qaeda leader Osama bin Laden was hiding in this deeply conservative Islamic region.

Western Pakistan's Islamic schools, or madrassas, have long been notorious for their anti-Western, pro-jihad (Islamic holy war) teachings, and have been fertile recruiting grounds for Islamic terrorist groups. In Fall 2002, fundamentalist Islamic parties swept elections in the Northwest Frontier Province, which voted the following June to adopt Islamic law, called Shariah. There are ongoing difficulties as the central government tries to reign in the terror-related activities of this "wild west."

The Least You Need to Know

- ◆ Rapid population growth means that South Asia will distance itself from East Asia as the world's most populous region.

- ◆ The world's largest democracy, India, has one sixth of the world's people.

- ◆ Subject to devastating monsoon rains and cyclones, Bangladesh seems to be on nature's hit list.

- ◆ Nepal's 29,028-foot Mount Everest is the world's highest mountain.

Chapter 24

Southeast Asia: Spice Islands and Golden Pagodas

In This Chapter

- ◆ Indochina or Southeast Asia?
- ◆ Peninsulas, archipelagoes, deltas, and jungles
- ◆ Southeast Asia's cities and people

Southeast Asia has long been thought of as a remote and exotic region—a land of beautiful people, spectacular temples, spice islands, and tropical forests filled with rare wildlife. Behind all the magic, however, lies a tortured political history.

Much of that history has been determined by geography. Southeast Asia is squeezed between the two Asian giants, China and India, which have largely shaped the region's culture. Its natural resources and strategic location made Southeast Asia a magnet early on for European colonization. (It was to reach China and the "spice islands" of Southeast Asia that Columbus sailed west from Spain and bumped into America.) In the decades after World War II, the region's struggle to put its colonial past behind it all too often resulted in war, revolution, more foreign intervention, genocide, repression, and lingering poverty.

What's in a (Place) Name?

The name Southeast Asia refers to the two peninsulas and two archipelagos that occupy the southeastern corner of the Asian continent—a region roughly bounded by China on the north; India, Bangladesh, and the Bay of Bengal on the west; the Indian Ocean on the south; and the Pacific Ocean on the east. What we now call Southeast Asia used to be referred to as the "East Indies" and sometimes "Indochina," although the latter term technically refers only to part of the region.

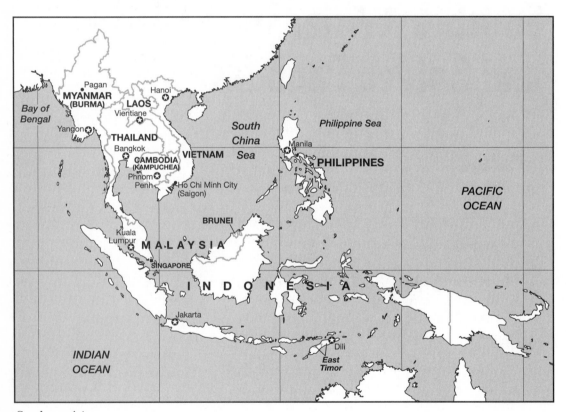

Southeast Asia.

The Indochinese peninsula is the larger of the two peninsulas that make up the Southeast Asian mainland. It consists of the three countries that were once part of French Indochina—Vietnam, Laos, and Cambodia—as well as most of Thailand (the only nation in the region that was never a colony), and part of the former British colony of Burma, now known as Myanmar. The region's second peninsula is the Malay peninsula, which stretches south from the Indochinese peninsula much like the tail of a kite. It includes parts of Myanmar and Thailand, the peninsular part of Malaysia, and at its tip, the city-state of Singapore.

South and east of the peninsulas are two archipelagos—the Indonesian and Philippine. The country of Indonesia occupies most of the Indonesian peninsula, except for the northwestern third of the island of Borneo, which is shared by Malaysia and the tiny state of Brunei, and the eastern half of Timor, which is now the new state of East Timor. The Philippine archipelago is occupied entirely by the Philippines.

Southeast Asia has given rise to many great civilizations and mighty kingdoms. Among the most splendid was the powerful Khmer empire, centered in the city of Angkor in present-day Cambodia. Southeast Asia's other great complex of ancient ruins, located in modern-day Myanmar (Burma), is the temple-city of Bagan, known as the city of four million *pagodas*.

def•i•ni•tion

The term **pagoda** refers to a variety of tower-like religious structures that are usually part of Buddhist temple or monastery complexes. The characteristic stupa of Southeast Asia is a type of pagoda shaped like a Hershey's kiss and often sheathed in gold. Stupas usually house some revered relic, such as a hair from the Buddha.

The Peninsular Countries

Peninsular Southeast Asia contains five countries and half of a sixth (Malaysia, which is split between the Malay peninsula and the island of Borneo).

Myanmar

Better known as Burma, Myanmar was renamed by the ruling military junta in 1989. When Burma was renamed Myanmar, its largest city, Rangoon (population: nearly 5 million), became Yangon. In 2005, the national capital was moved from Yangon to Naypyidaw. Diamond-shaped Myanmar occupies the northwestern portion of the Indochinese peninsula, with a narrow stretch of land jutting southward into the Malay peninsula. Bordered by Bangladesh, India, China, Laos, and Thailand, Myanmar is roughly the size of Texas.

Burma was colonized by the British in the nineteenth century and incorporated into British India. It became a separate, self-governing colony in 1937, only to be invaded by Japan in 1941. Burma was the scene of some of fiercest battles of World War II, many of them occurring along the famous Burma Road, the Allies' lifeline to China.

Burma gained its independence from Britain in 1948, and has had a troubled existence ever since. A 1962 coup established a socialist military regime that continues to suppress dissent and restrict basic freedoms.

Myanmar was once Southeast Asia's major rice exporter. Although the country remains heavily agricultural, it now barely feeds itself. Despite Myanmar's agricultural and mineral potential, its 50 million people remain among the world's poorest (per capita GDP $1,700).

Thailand

Thailand is located in the heart of mainland Southeast Asia, between Burma and the former French Indochina. Like Myanmar, Thailand has a thin, peninsular outcropping that reaches southward into the Malay Peninsula (called a proruption). Because of its unique configuration and position, Thailand has access to both the Adaman Sea and Indian Ocean on the west, and the Gulf of Thailand and South China Sea on the east. The country is about the size of France and has a population of nearly 65 million. The Thai capital, Bangkok, is known as the "Venice of the East" for its great waterway, the Chao Phraya River, and its many canals. This frenetic city is home to 6.6 million people (up to 15 million in its metropolitan area).

Since 1932, the country has been a constitutional monarchy. During World War II, Thailand was governed by a pro-Japanese puppet government and so was spared the full brutality of Japanese occupation. After the war, Thailand had its share of coups, attempted coups, and military governments. Thailand's diversified, free-market economy is one of the strongest in the region. Between 1985 and 1995, Thailand enjoyed one of the world's highest economic growth rates, until the Thai financial crisis of 1997 caused its currency to plummet and sent shock waves through markets and economies around the world. Since then, the Thai economy has been making a comeback, thanks mostly to strong exports. Thailand's major industries include textiles, food processing, electrical appliances, computer hardware, and mining (Thailand is the world's leading producer of tungsten and third largest tin producer). Tourism is a major money-earner, and although agriculture accounts for about 11 percent of Thailand's GDP, it employs more than half of its work force.

Laos

Landlocked and mountainous, especially in the north, Laos borders on Myanmar, Thailand, Cambodia, and Vietnam, as well as on China in the north. This densely forested country about the size of Utah supports (barely) a population of nearly 6 million. Its capital, Ventiane, is home to 200,000 people though the metro area population is over 700,000.

Thailand controlled Laos for nearly 200 years, until Laos became part of French Indochina in 1893 and did not gain its full independence until France withdrew from Indochina in 1954 following its defeat by the Vietnamese at the famous battle of Dienbienphu. Laos was inevitably drawn into the U.S.-Vietnam War and sustained heavy American bombing during the late 1960s and early 1970s. Within months of the fall of South Vietnam and Cambodia in 1975, the communist Pathet Lao party took over the country, forced the king to abdicate, and established the Lao People's Democratic Republic.

Despite attempts since 1986 to decentralize and privatize the economy, Laos is one of the poorest countries in a poor region (per capita GDP $2,000). Subsistence farming accounts for more than half of the country's GDP and employs more than 80 percent of the work force. Electricity is scarce, roads are few, and railroads nonexistent.

Cambodia

Bordered by the Gulf of Thailand to the west, Thailand to the west and north, Laos to the north, and Vietnam to the east and south, Cambodia physically resembles a bowl, with a large alluvial plain in the center surrounded by mountains. Even though Cambodia is smaller than Laos, its population of nearly 14 million is more than double that of its neighbor to the north.

Once the heartland of the great Khmer empire, with its spectacular twelfth-century temple-city complex at Angkor Wat, Cambodia became a French protectorate in 1863 and part of the union of French Indochina in 1893. Like Laos, Cambodia gained its full independence only after the end of the French-Indochina War in 1954, and Cambodia was drawn into the U.S.-Vietnam War and became a major battlefield in it.

Cambodia had the most horrific post-war experience. Communist Khmer Rouge forces under the monstrous Pol Pot took the Cambodian capital, Phnom Penh, in April 1975, and established the radical communist state of Democratic Kampuchea. During the Khmer Rouge reign of terror and the 1978 Vietnamese invasion that eventually ousted the regime, between 1.7 and 3 million people (out of a population of 7.3 million) were killed or died of starvation or overwork in the "killing fields" of Kampuchea.

Free elections in 1993 and the establishment of a constitutional monarchy under the country's longtime leader, Prince (now King) Sihanouk, seemed to signal an end to Cambodia's long nightmare, but the threat of political violence is never too far away.

Cambodia's modern history has left it with a tattered economy. Although growth rates have been remarkably strong in the first years of this century, Cambodia is still plagued by extreme poverty, especially in the countryside, and by the lack of skilled workers and basic infrastructure. Except for tiny East Timor, Cambodia has the lowest per capita GDP in the region. A strengthening tourism industry is a bright spot on the economic horizon.

Vietnam

Vietnam, the easternmost of the Southeast Asian peninsular countries, borders Laos and Cambodia to the west, China to the north, and the South China Sea to the east and south. This long, narrow country (31 miles wide at its narrowest) has a population of 85 million, making it the most populous country in peninsular Southeast Asia.

China ruled northern Vietnam for 1,000 years until the tenth century C.E., when a native Vietnamese dynasty took over and began expanding to the south.

French rule in Vietnam lasted from 1887 until 1954, when communist forces under Ho Chi Minh defeated the French. The country was divided into two parts: communist North Vietnam, supported by China and the Soviet Union, and noncommunist South Vietnam, whose authoritarian regime was backed by the United States.

Full-scale war between north and south raged throughout the 1960s and into the 1970s. In an effort to shore up the South Vietnamese government, the United States became increasingly involved in the conflict. At the height of the war in the late 1960s, more than half a million U.S. troops were stationed in Vietnam. Facing mounting antiwar protests at home, U.S. president Richard M. Nixon agreed to a peace settlement in January 1973 and pulled out the last U.S. troops from Vietnam later that year. In April 1975, the South Vietnamese government collapsed and the two Vietnams were finally united under communist rule. Its involvement in the Vietnamese War cost the United States the lives of 58,000 soldiers and more than $140 billion. The Vietnamese, however, lost 1.3 million people in the struggle.

With the fall of South Vietnam, its capital, Saigon, ceased being a national capital and had its name changed to Ho Chi Minh City, in honor of the victorious North Vietnamese leader. But with a population of more than 6 million (over 9 million in its metropolitan area), the former Saigon remains Vietnam's largest and most prosperous city. Hanoi, the national capital (and former capital of North Vietnam and of French Indochina), is less than half the size of Ho Chi Minh City.

Vietnam has been adopting a free-market economy while maintaining its communist political system. The results have been most impressive in the agricultural sector, with Vietnam now the world's second biggest rice exporter. The industrial sector, however, continues to stagnate. Vietnam is still a very poor country struggling to overcome the economic effects of years of war and destruction. Full diplomatic relations between Vietnam and the United States were restored in 1995, and in 2000, the two countries signed a bilateral trade agreement that is expected to bring significant benefits to Vietnam's economy.

Malaysia

Malaysia is both a peninsular and an island nation. Most of the country's nearly 27 million people live in West Malaysia on the southern end of the Malay Peninsula. Also located here is Malaysia's booming cosmopolitan capital, Kuala Lumpur, home to some 1.4 million people (6.5 million in its metropolitan area). Larger but much less populous East Malaysia occupies the northwestern third of the island of Borneo across the South China Sea. Both parts consist of coastal plains and interior mountains and are densely forested.

Much of Malaysia's history and economic development has been shaped by its location on the strategic Strait of Malacca, which very early on attracted the attention of European merchants and colonizers. First came the Portuguese in 1511, then the Dutch in 1641, and finally the British in the late 1700s. From their trading settlements on the Malay Peninsula—Malacca, Penang, and Singapore—the British went on to establish protectorates over the various peninsular Malay sultanates in the nineteenth and early twentieth centuries. After World War II, the British-held peninsular territories were granted semi-autonomy as the Federation of Malaya, and full independence in 1957. In 1963, the Federation of Malaysia was formed by the union of Malaya, Singapore, and the former British colonies on northern Borneo, Sabah and

Sarawak (now the two states that make up East Malaysia). Two years later, Singapore pulled out of the federation and became an independent country.

Over the last 30 or so years, Malaysia has transformed itself from a producer of raw materials (rubber, tin, timber, petroleum, and natural gas) to a country with a diversified and prosperous economy, driven largely by exports of manufactured goods, particularly electronics. Aside from Singapore and Brunei, whose economies are in a different league from those of their neighbors, Malaysia has the highest per capita GDP in Southeast Asia ($12,000).

Island Countries

Southeast Asia has thousands of islands, including some of the world's largest and some of the most beautiful. The island nations are Indonesia, Singapore, Brunei, East Timor, the Philippines, and half of Malaysia.

Indonesia

Indonesia's archipelago is the largest in the world, by number of islands, land area, and population. The country is made up of more than 17,500 islands straddling the equator; about 6,000 of them are inhabited. With a population of 231 million, Indonesia is the fourth most populous country in the world. Its capital, Jakarta, on the island of Java, is home to more than 8 million people (more than 23 million in its metropolitan area), making it the largest city in Southeast Asia.

The fabled Spice Islands of Indonesia have a very long and rich history. Buddhist and Hindu empires and Muslim sultanates flourished here long before the arrival of the Portuguese in the early sixteenth century. The Dutch followed later in the century and gradually extended their control over all the islands except part of one: the east side of tiny Timor remained under Portuguese control until 1975.

GeoRecord

Indonesia's thousands of islands include three of the world's largest. In this list of the largest islands in the world, square mileage is shown in parentheses and an asterisk (*) signifies an Indonesian island:

1. Greenland (840,004); 2. New Guinea* (309,000); 3. Borneo* (287,300); 4. Madagascar (226,658); 5. Baffin Island (195,928); 6. Sumatra* (182,860).

(New Guinea and Borneo are only partially occupied by Indonesia.)

The Dutch developed the Netherlands East Indies into one of world's most valuable colonial properties. During World War II, Indonesia endured a brutal Japanese occupation and afterward was briefly returned to Dutch control. After a bitter four-year war, Indonesia achieved its independence from the Netherlands in 1949. The Dutch colony in Western New Guinea became part of Indonesia in 1963.

Much of Indonesia's post-independence history has been marked by separatist rebellions and authoritarian rule, first under founding president Sukarno and from 1967 through 1998, under his successor General Suharto. In 1965, a coup attempt against "president for life" Sukarno prompted a vicious anticommunist purge in which an estimated 300,000 people were killed. Although free and fair national elections were held in Indonesia in 1999, the easing of authoritarian rule triggered violent separatist and sectarian uprisings in several parts of the country.

Indonesia has significant oil and gas reserves, abundant minerals and timber, and a developing manufacturing base. Bali and many of its other islands are tourist paradises. Yet Indonesia's economy is bogged down by lingering political instability, corruption, ethnic and regional conflicts, and threats of terrorism.

 **GeoRecord**

The world's largest Islamic country is not in the Middle East or in Northern Africa. It's Indonesia, with more than 190 million Muslims.

Brunei

Located on the northwest coast of the island of Borneo, squeezed between the two East Malaysian states of Sabah and Sarawak, Brunei is a peanut of a country, measuring a little more than 2,200 square miles. The population is mite-size also (about 374,000 people).

The Sultanate of Brunei had its first golden age about 500 years ago when it controlled all of Borneo and part of the Philippines. After a long period of decline beginning in the 1600s, Brunei became a British protectorate in 1888. Self-rule was restored in 1959, and the small country became fully independent in 1984.

Brunei might be very small, but it is also very rich. Little Brunei is the third largest oil producer in Southeast Asia and the fourth largest producer of natural gas in the world. Although down considerably from its 1980 peak, Brunei's per capita GDP ($23,600 in 2003) is second only to Singapore's in the region and far above that of most developing countries.

A hereditary constitutional sultanate, Brunei has been ruled by its current sultan, Haji Hassanal Bolkiah, since 1967. Oil revenues are used to subsidize education, health care, and even food and housing for the people of Brunei. The Brunei government also invests part of the country's oil revenues abroad as insurance against the day when the wells run dry.

Terra-Trivia

Scattered throughout the South China Sea are hundreds of tiny islands including two groups known as the Spratly Islands and the Parcel Islands. Although most of these islands are essentially uninhabited reef outcrops, China, Taiwan, Vietnam, Malaysia, the Philippines, and Brunei are all adamant about claiming them. Why? Whoever owns the islands of the South China Sea also owns the oil that lies beneath them.

East Timor

Southeast Asia's (and the world's) newest independent country is the former Portuguese colony of East Timor. This tiny (5,743-square-mile) nation of just under a million people is located on the island of Timor, one of the Lesser Sunda Islands at the eastern end of the Indonesian archipelago. In addition to the eastern half of the island, East Timor also occupies a small chunk of the northwestern coast and the two offshore islands of Pulau Atauro and Pulau Jaco. East Timor's capital, Dili, has a population of about 150,000.

The Portuguese began colonizing the island in the mid-seventeenth century. (The western half was ceded to the Dutch in the next century.) Portugal withdrew from its remaining colonies, including East Timor, in 1975. The East Timorese declared their independence on November 28, 1975, but 10 days later Indonesia invaded the country and incorporated it as a province. Over the next 20 years, as many as 250,000 people are thought to have died in the violent "pacification" campaign that followed Indonesia's illegal takeover.

In August 1999, a referendum supervised by the United Nations was held in East Timor and the results were overwhelmingly in favor of independence. That triggered another wave of violence and killings on the part of anti-independence paramilitary forces backed by the Indonesian army. Some 200,000 East Timorese fled to the western side of the island as their country was reduced to ruins. Finally, under the auspices of a United Nations Transitional Administration, presidential elections were held in April 2002 and East Timor became fully independent the following month.

Much of East Timor and its economic infrastructure was destroyed in the violence following the 1999 independence vote. A major infusion of international assistance has helped repair some of the damage, and most of the people who fled the violence have returned. One bright spot for the future is East Timor's untapped offshore oil reserves. But for now, the country has the unhappy distinction of having one of the world's lowest per capita GDP, a meager $800 in 2005.

The Philippines

Southeast Asia's other big archipelago is the Philippine Archipelago, a group of more than 7,100 volcanic islands off the east coast of Vietnam between the South China Sea and the Philippine Sea. The larger islands have mostly mountainous interiors surrounded by coastal lowlands. The entire group of islands is occupied by the nation of the Philippines.

Ferdinand Magellan claimed the islands for Spain in 1521 and for the next 377 years, they remained Spanish territory. After its defeat in the Spanish American War in 1898, Spain ceded the Philippines to the United States. In 1935, the country became a self-governing commonwealth. Japan invaded the Philippines in December 1941. Fulfilling his famous "I shall return" promise, General Douglas MacArthur landed in the Philippines on October 20, 1944, and after nearly a year of fierce and destructive fighting, it was Japan's turn to surrender. Despite the enormous devastation it had endured, the Philippines became independent in 1946.

After a series of orderly presidential successions during the post-war reconstruction period, Ferdinand Marcos came to power in 1965 and proceeded to rule despotically and corruptly for the next 21 years. The murder of opposition leader Benigno Aquino in 1983 sparked a "people power" uprising that eventually led to the ousting of Marcos in 1986 and the installation of Aquino's widow, Corazon, as the first post-Marcos president. Despite a number of attempted coups during Aquino's term, presidential successions have been more or less orderly in recent years. But as elsewhere in the region, corruption is a problem and the threat of instability persists. The country is also struggling to contain communist guerillas, Muslim separatist insurgencies in its southern islands, as well the growing threat of al-Qaeda-like terrorism.

Agriculture and light industries (mainly textiles, food processing, and electronics assembly) are the mainstays of the Philippines' economy. The country also has abundant mineral deposits, as well as significant natural gas, geothermal, hydropower, and coal reserves. Another asset is the high literacy rate (about 92 percent) of the

Philippines' 85 million people. Among the problems holding the economy back are corruption, political instability, the threat of insurgencies and terrorism, inadequate infrastructure, high levels of government debt, and a weakening currency.

Physical Facts

The region of Southeast Asia straddles the equator and is located almost entirely within the tropics (only the northern portion of Myanmar peeks above the Tropic of Cancer). The typical Southeast Asian terrain consists of a mountainous interior with fertile coastal plains and river lowlands.

Most of Southeast Asia is hot and wet. As is typical in tropical monsoonal climates, rainfall is heavy over the entire region between May and October, while northern areas receive considerably less rainfall between November and April. Tropical rain forests cover much of the region. Throughout Southeast Asia, the main crop is rice; other important crops include rubber, tea, spices, and coconuts.

Indonesia and the Philippines are Pacific Ring of Fire countries, which means that they are subject to destructive earthquakes and volcanic eruptions. Despite the risks, people have been drawn to these islands in great numbers over the centuries because of the fertility of their volcanic soils and their strategic trading location.

The Waters

Peninsular Southeast Asia is drained by several large and regionally important rivers. The Mekong River is Asia's sixth longest river and the thirteenth longest in the world. From its source in Tibet, the Mekong flows southward through China, along the Thai–Laotian border, through Cambodia, and on to its huge delta in southern Vietnam. Two of the region's most important rivers flow through Myanmar. Dominating the entire central portion of the country is the great Irrawaddy River, Myanmar's revered lifeline. Along its banks are the nation's two largest cities and biggest population concentrations. Originating deep in the heart of Chinese Tibet, the Salween River flows through eastern Myanmar, forming part of the Thai-Myanmar border. Thailand's Chao Phraya, famous for the golden pagodas that line its banks, flows down the western part of the country and through the capital city of Bangkok.

In the west, two arms of the Indian Ocean abut Southeast Asia—the Bay of Bengal and the Andaman Sea. The peninsular mainland is separated from the islands by the famous spice trade route, the Strait of Malacca. Amid the Indonesian islands are the Java Sea and the Banda Sea, while the Timor Sea and the Arafura Sea separate Indonesia from Australia. Between Indonesia and the Philippines is the Celebes Sea. East of Vietnam lies the South China Sea; east of the Philippines is the Philippine Sea.

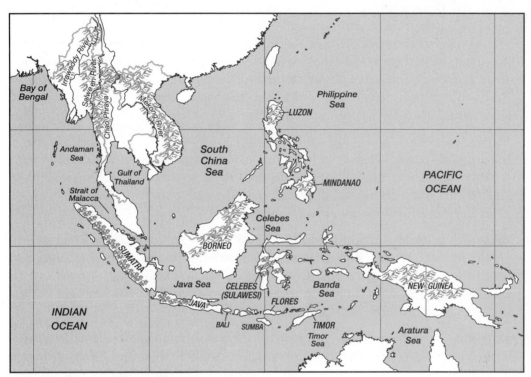

Southeast Asia—physical features.

The Islands

The region's largest island is New Guinea. Although Indonesia occupies half of it, geographically New Guinea is considered to be part of Melanesian group of islands in Oceania. The second largest island in Southeast Asia, and the largest in the Indonesian archipelago, is Borneo, which is shared by Indonesia, Malaysia, and Brunei. Borneo, Sumatra, Java, and Sulawesi (Celebes) make up the Greater Sunda Islands. Java is Indonesia's core and the site of its capital, Jakarta.

East of Java are the Lesser Sunda Islands, extending from Bali to Timor. Between Sulawesi and New Guinea are the Moluccas, better known as the Spice Islands.

North of Indonesia are the thousands of islands that make up the Philippine archipelago. Luzon in the north is the main population and industrial center and site of the Philippine capital, Manila. The Philippines' second largest island is Mindanao in the south, home to most of the country's Muslim population.

The Mix of Peoples

The mountainous nature of much of Southeast Asia has helped preserve the distinctiveness of many ethnic groups. On the mainland, the major groups roughly coincide with national borders: Burman peoples in Myanmar, Thai in Thailand, Lao in Laos, Khmer in Cambodia, Malay and Chinese in Malaysia, and Vietnamese in Vietnam. But the divisions aren't always this clear-cut, and grievances and aspirations of ethnic minorities continue to be a source of actual or potential conflict throughout the region.

The islands of Indonesia and the Philippines contain hundreds of smaller ethnic groups, but in general, the Indonesian mix consists of Javanese (45 percent) and smaller numbers of Sundanese, Madurese, coastal Malays, and in Indonesian New Guinea, Papuans. Filipinos are predominantly Malay.

The region's languages basically parallel its ethnic patterns. Some French lingers in the three former French Indochinese countries, although even in Vietnam English is becoming the more popular second language. English is spoken by business and government elites in many Southeast Asian nations and is an official language of the Philippines along with the national language, Pilipino.

Southeast Asia is a land of many religions. The peninsular mainland is predominantly Theravada Buddhist; Indonesia is mainly Muslim, while Roman Catholicism prevails in the Philippines, and the entire region has pockets of Hindu, Mahayana Buddhism, traditional; and other faiths.

The Least You Need to Know

- Mainland Southeast Asia is a peninsular extension of Asia; the region also contains thousands of large and small islands.

- Southeast Asia is heavily influenced by its two large neighbors, India and China.

- Southeast Asia is hot and wet, with rain forests and mountainous interiors.

- Southeast Asia has many ethnic groups, numerous languages, and the world's largest concentration of Theravada Buddhists.

Chapter 25

The Vast Pacific

In This Chapter

- Surveying the world's oceans
- Probing the Pacific and its peoples
- Sorting through the -nesias: Mela-, Micro-, and Poly-

The Pacific region consists mainly of ocean with a whole lot of islands. It has no continental landmass, and most of its islands are quite small.

The Pacific is the deepest ocean, with an average depth of 12,925 feet. It leads all other oceans with 16 trenches more than 20,000 feet deep. The Pacific's trenches are also much deeper than any others. The Pacific has three deeper than 30,000 feet: the Philippine Trench, just shy of 33,000 feet; the Tonga Trench, about 35,400 feet; and the deepest spot on earth, the Mariana Trench's Challenger Deep, at 35,840 feet. The Atlantic's deepest trench, the Puerto Rican Trench, measures about 28,200 feet.

The Pacific Ocean has the largest surface area, measuring more than 64 million square miles and is almost as large as the other three oceans combined.

The Pacific

The Pacific Ocean's maximum length is 9,600 miles from north to south and its maximum width is 11,000 miles from west to east at the equator. The Pacific covers one third the earth's surface and holds more than half of its free water.

The Pacific Ocean is bounded on the north by Russia and Alaska, on the east by the Americas, on the south by Antarctica, and on the west by Asia and Australia. The Pacific connects to the Arctic Ocean via the Bering Strait; to the Atlantic via the Drake Passage, the Straits of Magellan, and the Panama Canal; and to the Indian Ocean via various passages in the Indonesian Archipelago and between Australia and Antarctica.

The Pacific Ocean may be vast, but the Pacific region is very small in terms of overall landmass. The region consists of mostly tiny Pacific islands located south of the Tropic of Cancer and known collectively as Oceania. Although Oceania comprises as many as 30,000 islands, all of them, even including the regional giant, New Guinea, account for only about two thirds of 1 percent of the Pacific realm. Other Pacific islands and island groups, such as Japan, Taiwan, the Philippines, and the Indonesian archipelago, are not usually considered part of Oceania and are covered elsewhere in this book. New Zealand and even Australia are sometimes included in Oceania.

The Not-So-Peaceful Giant

The name Pacific means "peaceful," but the Pacific isn't really very peaceful. From Antarctica, running north for about 6,000 miles, is the East Pacific Rise, which thrusts up about 7,000 feet. Along this rise, molten rock swells up from the earth's mantle and adds crust to the ocean plates on both sides. The plates are forced apart, causing them to collide with the continental plates all around them. The tremendous pressure generated by this process forces the continental plates to turn upward as mountains and the ocean plates to turn downward into deep trenches.

Island Origins

Tectonic pressures along the Pacific's perimeter have created the violent seismic and volcanic zone known as the Ring of Fire. Hot spots in the earth's mantle can cause molten rock to break through the surface. When this rock is beneath the ocean, a seamount (volcanic island) is born. Thousands of the "dots" that have appeared all over the Pacific, particularly in the southwest, were formed this way. Native inhabitants call these formations "high" islands.

Thousands more of the Pacific dots are so-called "low," or coral islands. Coral islands and reefs have been described as the earth's most complex ecosystems. They occur only in warm, sunlit water and result from thousands of live marine animals clustering in vast colonies. The skeletal deposits that remain after these creatures die eventually layer themselves and can reach a height of 300 or 400 feet.

Currents: Air and Sea

To best understand the Pacific Ocean's wind patterns, start at the equator and work north and south from there. Between 10 degrees north and 10 degrees south are the doldrums, a low-pressure area of light breezes. At 30 degrees north and 30 degrees south are the horse latitudes, belts of high pressure. The winds blowing from these high-pressure areas toward the low-pressure doldrums are deflected by the earth's rotation and become the tropical easterlies known as trade winds. Between 30 degrees north and 60 degrees north and between 30 degrees south and 60 degrees south, winds blow from the horse latitudes toward the poles and again are deflected by the earth's rotation to become the westerlies. The polar regions are high-pressure areas; the deflected winds blowing from them become the polar easterlies.

Although ocean currents in the Pacific are relatively complex, generally, north of the equator, the currents form a huge oval that moves clockwise. A similar oval current forms south of the equator, in which the current moves counterclockwise. Between the two ovals and running along the equator is a west-to-east movement of water, called the equatorial counter current. Sometimes, these regular patterns may be disrupted by a periodic phenomenon called El Niño. Every two to seven years the Pacific trade winds slacken or reverse direction, causing surface water temperatures to rise. This El Niño Effect causes weather havoc—including flooding, hurricanes, and droughts across the globe.

Countries in an Endless Ocean

New Guinea and Oceania combined have almost 9.5 million people and encompass some 30,000 islands, which are divided into three major groupings—Melanesia, Micronesia, and Polynesia—along with a few scattered outliers. The groupings include 13 independent countries; one Indonesian province; one United States state; five other United States affiliates; one British, four French, one Chilean, and three New Zealand territories; and a part of Ecuador.

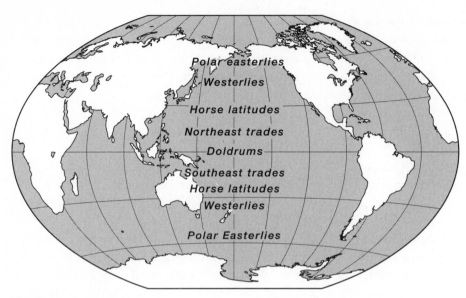

Pacific wind patterns.

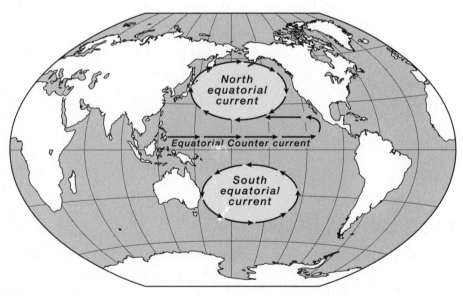

Pacific currents.

Melanesia

The term Melanesia comes from two Greek words meaning "black islands"—a reference to the complexion of most Melanesians. Melanesia has nearly 80 percent of the region's population (about 7 million people) and 93 percent of its land area. The reason that Melanesia looms so large is that it includes the large and heavily populated nation of Papua New Guinea.

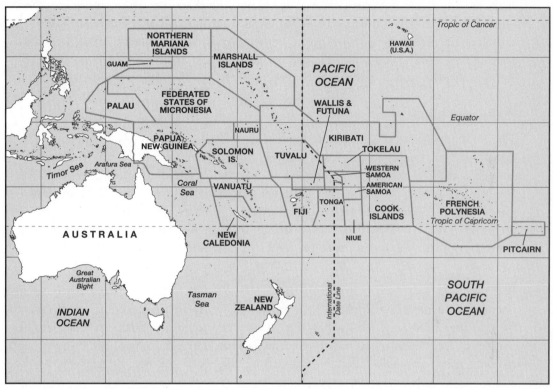

Islands of the Pacific.

Papua New Guinea

The world's second-largest island, New Guinea, is divided almost evenly between Indonesia's West Papua (formerly Irian Jaya) province on the western half of the island and the Melanesian country of Papua New Guinea on the eastern half. In addition to its share of the big island, Papua New Guinea also encompasses some 600 smaller islands, including the Admiralty Islands, the Bismarck Archipelago, and Bougainville. Papua New Guinea is home to nearly 5.9 million people. Its capital, Port Moresby, has a population of almost 325,000, making it one of the largest cities in the region.

Papua New Guinea has a hot and humid tropical climate and dense forests. Rainfall varies with topography, with some places receiving more than 230 inches per year.

The interior of the country is extremely rugged and the tribespeople who live there, the Highlanders, were once among the most isolated people on Earth.

Papua New Guinea became a fully independent country in 1975. The country's more than 700 Papuan and Melanesian tribes speak more than 700 languages. Papua New Guinea has abundant mineral and other natural resources, but it has been slow to exploit them. Tribal people fish, hunt, and engage in subsistence farming. In coastal areas, cocoa, coffee, and copra are grown for export. This is one of the most rural countries in the world, with only 18 percent of the population living in cities.

The Solomon Islands

The Solomon Islands became independent in 1978. The country is composed of 10 major volcanic, deeply forested islands (including Guadalcanal), 30 small isles, and countless atolls. The approximately 495,000 Solomon Islanders speak an estimated 120 local languages. A Melanesian-English pidgin is used to bridge the many language divides. The official language, English, is spoken by very few. Despite the presence of rich mineral resources, farming, fishing, and forestry are the economic mainstays of this chaotic country plagued by crime and intertribal strife.

Vanuatu

Formerly known as New Hebrides, Vanuatu was jointly administered by the British and French from 1906 until 1980, when the country became independent. About 211,000 people inhabit Vanuatu's 70 mostly volcanic islands. The economy of the resource-poor nation is based on small-scale farming, fishing, offshore banking, and tourism.

New Caledonia

A French possession since 1853, New Caledonia became an overseas territory of France in 1956. Its population of nearly 237,000 is divided between Melanesians (about 43 percent) and French or other Europeans (37 percent). New Caledonians voted against independence in 1998. The country's economy depends on tourism and nickel mining. (New Caledonia has about 25 percent of the world's nickel supply.)

Fiji

After nearly a century of British colonial rule, Fiji gained its independence in 1970. The country consists of 300 islands (only 100 are inhabited) with a population of about 905,000. More than half the people are Fijian (mainly Melanesian); about 44 percent are Indian, descendents of contract laborers brought to work on Fiji's plantations in the late nineteenth century. Friction between the two groups has led to coups and political turmoil in recent years. Fiji's capital, Suva, has about 177,000 inhabitants. English is the official language, but almost everyone speaks Fijian; Hindustani is also widely spoken. Even though about 70 percent of its work force is engaged in agriculture, Fiji has one of the most developed economies in the region; tourism and sugar are the leading money earners.

Micronesia

The term Micronesia comes from Greek words meaning "small islands." The sub-region's more than 2,000 islands cover only 1,000 square miles and support some 580,000 people. Although tiny coral atolls are the norm, the bulk of the population lives on the larger high islands.

Palau

The Palau islands, the westernmost archipelago of the Caroline chain, were controlled by Spain, Germany, Japan, and finally the United States until they became independent in 1994. They are still linked to the United States by a Compact of Free Association. Palau consists of more than 300 islands (half in Melanesia and half in Micronesia) divided into six groups. More than half of Palau's 20,000 people live in the capital, Koror. Tourism, farming, and fishing are the main ingredients of Palau's economy.

The Federated States of Micronesia

Formerly a United Nations Trust Territory administered by the United States, the Federated States of Micronesia became independent in 1986 but retained its ties to the United States under a Compact of Free Association. The country's four island groups contain more than 600 islands (many uninhabited), ranging from high volcanic islands to low coral atolls. Its population numbers 136,000 and its economy is heavily dependent on United States aid.

Nauru

With some 13,000 people packed on one 8-square-mile island, Nauru has a population density of about 1,600 people per square mile. Independence was achieved in 1968. Phosphate mining made Nauru relatively prosperous, but supplies are almost exhausted and the island has very few resources to compensate for the loss of the phosphate industry. Even fresh water has to be imported.

The Northern Mariana Islands

The Northern Mariana Islands are officially a "commonwealth in political union with the United States." The Marianas are a group of 14 islands, including Saipan, Tinian, and Rota, with a population of about 80,000. The chief industries are tourism, small-scale farming, and garment manufacturing. This last industry has come under severe criticism for using imported Chinese labor to produce "made in the U.S.A. garments" under deplorable sweatshop conditions.

Guam

Large by Micronesian standards, the island of Guam measures 209 square miles and is home to nearly 170,000 people. Guam was ceded to the United States by Spain in 1898 and is today an unincorporated territory of the United States. About one third of Guam's land is owned by the United States military, which maintains a highly strategic base on the island. Military expenditures and tourism are the mainstays of Guam's economy.

The Marshall Islands

Like Micronesia, the Marshall Islands became independent in 1986 under a Compact of Free Association with the United States. The country consists of two parallel chains of atolls and islands with about 62,000 inhabitants. The atolls include Kwajalein (the world's largest atoll, with 90 islets surrounding a 650-square-mile lagoon) and the former nuclear test atolls of Bikini and Enewetak. The capital, Majuro, is home to more than a third of the country's total population. The Marshall Islands are highly dependent on United States aid and continue to seek compensation for the damage done to Bikini, Enewetak, and other atolls by the U.S. nuclear testing program.

Kiribati

The Gilbert Islands, a group of 33 coral atolls straddling the equator and the international dateline, gained its independence from the United Kingdom in 1979 and changed its name to Kiribati. Its population is about 99,000. Its capital, South Tarawa, on the Tarawa Atoll, is remembered as the site of a horrific World War II battle between the United States Marines and Japanese forces.

Polynesia

The name Polynesia stems from a Greek word meaning "many islands." Indeed, Polynesia consists of many thousands of coral and volcanic islands in what is by far the most far-flung of the three major Pacific island groupings. Nearly 2 million people live in the various political configurations that make up Polynesia.

Tuvalu

Once part of the same British colony as the mostly Micronesian Gilbert Islands (now Kiribati), the mostly Polynesian Ellice Islands decided to become the separate colony of Tuvalu in 1974. Four years later, Tuvalu became independent. Tuvalu's 11,000 inhabitants are packed on 10 square miles of land over nine atolls in a 360-mile chain. Global warming, which could cause sea levels to rise, is a nasty phrase in this tiny, remote, and very low-lying country. Tuvalu's highest elevation is only 15 feet.

Tonga

Tonga is made up of 170 coral and volcanic islands in a 289-mile-long north-south chain. Only 36 of the islands are inhabited by about 106,000 Tongans. Named the Friendly Islands by Captain Cook, Tonga was a British protectorate from 1900 to 1970, when it gained its independence as a hereditary constitutional monarchy (the only monarchy left in Oceania). The country's fairly stable economy depends on agriculture, tourism, and remittances from Tongans working abroad. Tonga has recently been wracked with internal disorder as pro-democracy rioters battle against the Tongan monarchy.

New Zealand Territories

Three Polynesian island groups have different levels of association with New Zealand. The Cook Islands (named for, you guessed it, Captain Cook) were a British protectorate from 1888 to 1900, when New Zealand took over. In 1965, the Cook Islands became a self-governing democracy in free association with New Zealand. Most of the 18,700 inhabitants live on the eight elevated volcanic islands of the southern Cook Islands; the northern atolls are sparsely populated. The remote island of Niue (called "The Rock" by its inhabitants), one of the world's largest coral islands but with a population of only 2,100 and falling, is also a self-governing democracy in free association with New Zealand. The three coral isles that make up Tokelau cover four square miles and support 1,400 people. Tokelau is a self-administering territory of New Zealand, but is moving toward free association.

French Territories

French Polynesia consists of five archipelagos acquired by France in the nineteenth century: the Society, Tuamotu, Gambier, Austral, and Marquesas Islands. The islands became an overseas territory of France in 1946, with Papeete, on the island of Tahiti, as the capital. This widely spaced collection of islands is home to more than 260,000 people and includes some of the most beautiful getaway spots on Earth. The best-known islands are Tahiti (immortalized by the great French post-Impressionist painter, Paul Gauguin), Moorea, and Bora Bora. Tourism, not surprisingly, is the main industry, followed by pearl farming, agriculture, and deep-sea commercial fishing. The island groups of Wallis and Futuna became a French overseas territory in 1959. Most of the nearly 16,000 inhabitants earn their living by farming and fishing.

Samoa

Samoa (formerly German Samoa and Western Samoa) was German territory from 1899 to 1914, when it was taken over by New Zealand. In 1962, Western Samoa became the first Polynesian nation to regain its independence in the twentieth century. The "Western" part of its name was dropped in 1997. Samoa consists of two large volcanic islands, Savai'i and Upolu, and seven smaller islands that support a declining population of about 185,000. Agriculture, development aid from New Zealand and Australia, and remittances from Samoans working abroad are the mainstays of the economy.

American Samoa

The seven eastern islands of the Samoan chain are an unincorporated territory of the United States. American Samoa's largest island is Tutuila, site of the capital, Pago Pago, one of the finest harbors in the South Pacific. The territory has a population of about 65,000. The economy of American Samoa depends heavily on tuna fishing and canning and on U.S. government subsidies.

Hawaii

Unlike the other United States possessions in the Pacific, Hawaii is a full-fledged state. The Hawaiian islands became a territory of the United States in 1900 and the fiftieth state of the Union in 1959. The 1,523-mile-long Hawaiian island chain consists of eight main volcanic islands and 124 islets and reefs. The largest island is Hawaii, the "Big Island." The state's capital and, with 377,000 residents, its largest city is Honolulu, located on the island of Oahu. The lushest island is Kauai, home of Mount Waialeale, the world's wettest place with an average annual rainfall of 486 inches. The island of Molokai is perhaps still best known as the site of Father Damien's famed leper colony. Hawaii's population of 1.2 million is very diverse: 42 percent Asian, 24 percent white, 9 percent native Hawaiian or Pacific islander, 7 percent Hispanic, and 2 percent black.

Hawaii's earliest settlers were Polynesians, probably from the Marquesas Islands, who arrived in outrigger canoes beginning in around 300 C.E. Although Europeans began arriving in significant numbers in the 1790s, bringing with them diseases that decimated the native population, a native monarchy ruled until 1893, when government officials, sugar planters, and American businessmen engineered its overthrow. A republic was declared the next year, and in 1898, the islands were annexed by the United States.

Sugar production was introduced in the islands in the 1830s and soon replaced whaling as the primary industry. In 1898, the U.S. naval base at Pearl Harbor was opened, adding a major element to Hawaii's economy—military expenditures. Today, tourism has eclipsed both agriculture and military spending as the base of Hawaii's economy.

Miscellaneous Islands

Wake Island (north of the Marshalls) and the Midway Islands (west of Hawaii) are unincorporated territories of the United States and run by the military; the Midway Islands are a wildlife refuge run by the U.S. Fish and Wildlife Service. Two others are France's uninhabited Clipperton Island (west of Costa Rica) and the Galápagos Islands (west of and part of Ecuador), which are famous for gigantic tortoises. Easter Island lies roughly halfway between Chile, its owner, and French Polynesia. Easter Island is renowned for its mysterious huge-headed, long-eared, long-nosed stone statues.

The People of New Guinea and the Oceania Dots

Melanesians are believed to be the result of a mix between Papuan peoples and the black-skinned people of interior New Guinea, who then spread out to the islands between the equator and 20 degrees south latitude. More than 200 Malayo-Polynesian languages are now spoken in Melanesia, often several on the same island. Scattered throughout many of the southern Solomons and the northern Vanuatus are small-statured tribes who to this day practice polygamy and engaged in cannibalism in the no-too-distant past.

The Micronesians descend from a historic mixture of Asian, Indonesian, Filipino, and Papuan peoples. They are generally slender with straight hair and lighter skin than Melanesians. Micronesians were skilled seamen who navigated their craft between the islands to supplement their diets. The high islanders farmed, and the low islanders fished.

The Polynesians are thought to have come from the Malay Peninsula via Indonesia, probably in two major waves. The first wave settled in Samoa, Tonga, and Fiji; the second, a few hundred years later, migrated westward, northward, and southward (into New Zealand, whose Maori are thought to be of Polynesian descent). Polynesians have light brown skin and are known for their grace and beauty. The Polynesians were also renowned for their seamanship, both in small canoes for short inter-island trips and in large craft that carried them over vast distances to discover, explore, and settle new islands thousands of miles away.

The Least You Need to Know

- ◆ The size of the Pacific Ocean almost equals the Atlantic, Indian, and Arctic oceans combined. It covers one third of the entire earth's surface.

- ◆ The Pacific's entire rim, the Ring of Fire, has active volcanoes and is prone to earthquakes.

- ◆ The Pacific has almost 30,000 islands (mostly atolls) that are volcanic and coral.

- ◆ Oceania is divided into subregions called Melanesia, Micronesia, and Polynesia.

Chapter 26

The Frozen Poles

In This Chapter

- ◆ Identifying polar limits and who owns what
- ◆ Highlighting polar similarities and differences
- ◆ Keeping an eye on the ozone hole

This book typically defines regions in terms of both physical and cultural characteristics. Because the poles don't fit this mold, however, this chapter will primarily focus on physical features only.

Defining the Polar Regions

Determining the limits of the polar regions is not as clear-cut a proposition as you might expect. Some definitions are based on climatic conditions. For example, the Arctic and Antarctic regions can be defined as the parts of the globe in which the average temperature of the warmest month never exceeds 50 degrees F. Some geographers define the Arctic and Antarctic regions by vegetation: they are the parts of the globe above the latitude at which trees no longer grow (the tree line).

You might think that the logical boundary lines of the polar regions would be the Arctic and Antarctic circles, at about 66½ degrees north and south latitude, respectively. As tidy as those lines would be, they don't quite work.

The poles.

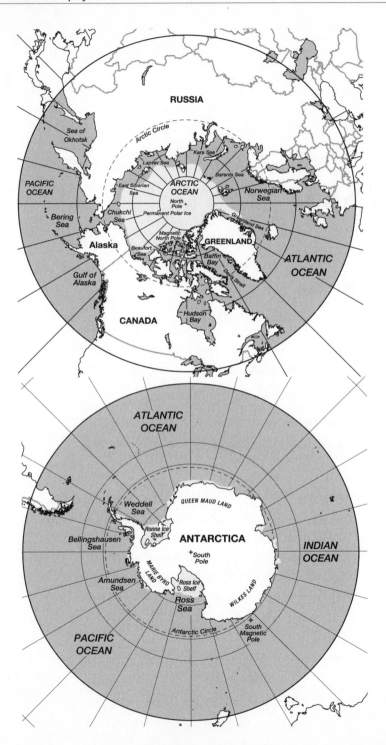

Although Antarctica is contained almost entirely within the Antarctic Circle, the continent peeks outside the circle in several locations. In the Eastern Hemisphere, parts of Wilkes Land and Enderby Land flirt with the Antarctic Circle, and bits of these areas poke out along the way. In the Western Hemisphere, Antarctica is contained entirely within the Antarctic Circle except for the northernmost portion of the Antarctic Peninsula.

Geographically Speaking

The earth's true, or geographic, poles are the ends of the axis around which the planet rotates. The magnetic poles are the ends of the axis of the earth's magnetic field. The magnetic poles don't coincide with the earth's true poles, and they aren't stationary either. The North and South magnetic poles are actually average positions around which the poles wander on a daily basis. The average positions also shift over time. Currently, the North Magnetic Pole is drifting across the Canadian Arctic. The South Magnetic Pole is off the coast of Wilkes Land, Antarctica.

The opposite situation exists in the Northern Hemisphere. The Arctic Circle encompasses parts of several other regions not considered part of the north polar region. A quick look at the map shows that Russia, Northern Europe, and North America all extend into the Arctic Circle. The north polar region includes everything north of these other landmasses, and therefore, contains no land itself, just water and ice.

Making National Claims

Because the North Pole is located in the heart of the Arctic Ocean, there is no land to claim in the north polar region. Even with the 200-mile exclusive economic zone provided by the Law of the Sea Convention, no country's jurisdiction reaches the North Pole. At the South Pole, however, the situation is not so simple.

Antarctica is land, and although it has no permanent inhabitants and no local government, it has not escaped political squabbling. Seven different countries make claims to portions of this continent. No nation claims the part of Antarctica called Marie Byrd Land; this is the place to go if you're looking to claim a chunk of a continent.

The pie-shaped national claims all originate at the South Pole. Most of the claims have some logic behind them. Australia (which claims two slices of Antarctica), New Zealand, Argentina, and Chile—the countries closest to Antarctica—all make claims that are southern extensions of their national territories. Norway also makes a (more dubious) claim to a wedge of Antarctica, based on the fact that a Norwegian, Roald Amundsen, was the first person to make it to the South Pole. Because Robert Scott, from the United Kingdom, was a close second, the British claim also has some rationale. The other claimant is France, and if you don't get the logic on this one, join the club.

All these claims are something of a moot point anyway because of the Antarctic Treaty System, which came into effect in 1961. This treaty banned new territorial claims and prescribed peaceful, scientific use of the continent. A more recent agreement, signed in 1991, prohibits commercial mineral extraction on Antarctica. You don't have to look any further than the Antarctic Peninsula to see the potential for conflict: three countries all claim the same area.

Looking at Polar Similarities and Differences

Even though the north and south polar regions are literally "poles apart," they do share certain similarities. Both are located at 90 degrees latitude (ignore the north and south issue). During winter, the sun never rises, and during the summer, the sun never sets, in either place. Both have lots of ice and snow, are extremely cold, and seals live on both poles. Neither has trees; and, if you stand on either pole, no matter which way you face, you can only travel in one direction (only south from the North Pole and vice versa). Plus, you can forget about using your compass in either place.

But before you begin to think that the poles are almost identical, take a moment to peruse the top nine differences between them. The poles are 180 degrees apart, on opposite sides of the globe—as far apart as you can get on Earth. No polar bears live on Antarctica; and, while the South Pole is on land, the North Pole is ice on water. Both are cold, but the South Pole is considerably colder than the North Pole. The South Pole is almost entirely covered in glacial ice, while the North Pole has very few glaciers. The South Pole has penguins, but the North Pole does not. Antarctica is mountainous, with one peak 16,000 feet high, while the North Pole is pretty much at sea level. Nations have made claims on the South Pole's territory and set up permanent scientific stations, but nobody has made any claims on the North Pole and only temporary floating stations are currently studying it.

The poles seem far away, but what happens to the polar regions has environmental repercussions for the entire planet.

The North Pole is in the Arctic Ocean. The smallest of the world's oceans, it's also the shallowest and, obviously, the world's northernmost ocean. Most of the Arctic

Ocean, including the North Pole, is always covered with ice. This ice is not a solid sheet; it's called pack ice—chunks of ice that slowly circulate about the pole.

Eye on the Environment

Another important similarity between the poles is that both are barometers of global warming. The discovery of open water at the North Pole in 2000 raised worldwide alarms. Although the alarms might have been overplayed—arctic ice floats on top of the Arctic Ocean and is in constant motion, often exposing areas of open water, especially in the summer—the extent and thickness of Arctic ice are clearly declining. According to some projections, by the middle of this century, the Arctic Ocean might be ice-free during the summer.

The situation in Antarctica is no less disturbing. The process known as "calving," in which large chunks of the Antarctic ice shelves break off at the fringes and float off into the sea as icebergs, is a normal one. But the rate of disintegration and the size of the resulting icebergs have been increasing alarmingly in recent years. Although the cause of this phenomenon is still a matter of some debate, the evidence pointing toward global warming cannot be ignored.

The South Pole is on the continent of Antarctica. Even though almost its entire surface is covered with continental glacial ice, Antarctica is solid land. Antarctica is the fifth largest of the earth's seven continents. It's smaller than Asia, Africa, and North and South America, and larger than Europe and Australia.

Polar Extremes

Plateau Station on Antarctica is the coldest place on Earth, with an annual average temperature of –70 degrees F. The single coldest temperature ever recorded was also on Antarctica, at Vostok Station, where the thermometer dipped to –129 degrees F. Antarctica is also one of the world's windiest places. At Australia's Commonwealth Bay Station, cold air constantly rushes down from the ice cap, and wind speeds have been recorded at more than 180 mph.

Gauging Arctic temperatures and winds is a little more difficult than in the Antarctic because no fixed, permanent stations exist. Although you would think that the coldest temperatures in the north would be at the North Pole, the coldest are farther south along the Arctic Circle. This phenomenon occurs because the North Pole is located over water, and the cold temperatures recorded along the Arctic Circle have been recorded over land.

Just outside the north polar region, at 76½ degrees north latitude, is Thule Air Base in Northern Greenland. Annual temperatures average about –11 degrees F—cold, but not approaching the –70 degrees F recorded at Antarctica. Verkhoyansk Station in Northern Siberia has recorded January temperatures as low as –90 degrees F—again, it's cold, but not as bad as at Vostok Station in Antarctica.

There are several reasons for the temperature differences between the poles. At the height of the south polar winter, around July 1, the earth is at its aphelion, or farthest point in its orbit from the sun. Although that's not a huge factor in temperature, it does mean that the South Pole is about three million miles farther from the sun during its winter than the North Pole is during its winter.

The windier conditions at the South Pole make the wind-chill a factor to be reckoned with. To demonstrate the impact of the wind-chill effect, if the recorded temperature were –45 degrees F (a warm day at Plateau Station, Antarctica), winds of 45 miles per hour would make the temperature feel like –125 degrees F.

Polar Animals and Glaciers

Aside from an occasional mite, there's no purely terrestrial life at the poles—all polar life is dependent on the seas. Under the ocean ice, algae and diatoms support small shrimplike creatures, called krill, that provide food for fish and larger aquatic life at the poles.

Penguins live only at the South Pole and are the only large animals that stick it out through the nasty Antarctic winter. Seals live on both poles. In the far north, polar bears as well as an occasional Inuit hunter have been known to venture out on the pack ice in search of seal meat. No such indigenous land predators exist on Antarctica to plague the penguins.

Only two remnants of the continental glaciers remain: one covers the island of Greenland, and the other covers Antarctica.

Continental glaciers form when climatic factors combine to allow a gradual buildup of snow and ice. When the accumulation of ice builds to a critical mass, the weight of the ice mass causes it to flow outward. As the glaciers advance, they grind and scrape the landscape with dramatic results.

Terra-Trivia

The Antarctic ice cap contains the world's largest supply of freshwater, approximately 70 percent of the total. More freshwater is contained in the continental glacier of Antarctica than the freshwater in all the world's rivers, lakes, atmosphere, groundwater, and other glaciers combined.

Either way, glaciers have not historically formed over the North Pole. The continental glaciers of the Northern Hemisphere formed in subpolar areas and then spread northward toward the pole and southward toward more temperate climes. The ice cap on Antarctica is not advancing because of the scant amount of snowfall the continent receives.

Night and Day

When the sun rises at either pole, it stays daylight for six months. When the sun finally sets, nighttime settles in for six months. The cause of this pattern is the $23\frac{1}{2}$-degree tilt of the earth on its axis. When it's summer in the Northern Hemisphere (March 21–September 23), the North Pole remains in the sunlight and the South Pole stays in the darkness. During the Southern Hemisphere's summer (September 23–March 21), the South Pole experiences constant daylight and the North Pole languishes beneath dark winter skies.

In the News: The Ozone Hole

The ozone layer in the upper atmosphere shields life on Earth from the harmful effects of the sun's ultraviolet radiation. Normally, the ozone layer covers and protects the entire earth, but in 1985 scientists noticed a hole forming in the ozone layer directly over Antarctica. Since then, scientific observations have confirmed a 15 to 70 percent decline in ozone levels over the South Pole every spring.

The depletion of the ozone layer is an unfortunate result of human activities. The release into the atmosphere of certain chemicals, particularly a family of chemicals called chlorofluorocarbons, or CFCs, has broken down atmospheric ozone and will continue to do so for some time despite successful international efforts to ban CFC use and production. Thanks to the circulation and concentration of cold air over Antarctica (a pattern called the Antarctic vortex) the problem is more severe over the South Pole. Although ozone depletion also occurs over the North Pole, the loss there has not been as great.

Because it's believed that excessive ultraviolet radiation can hinder plant development, Antarctica's ozone hole might be placing its basic food chain at risk. Leopard seals eat penguins and penguins eat krill and krill eat algae. If algae are affected, so, too, are the higher vertebrates on the chain. Because algae also absorb carbon dioxide and produce oxygen, their loss might also have additional negative effects on the atmosphere.

The Least You Need to Know

- The earth's true, or geographic, poles are the ends of the axis around which the planet rotates. The geographic poles do not coincide with the magnetic poles.

- Although both poles are extremely cold, the South Pole is colder.

- The North Pole is over the Arctic Ocean, and the South Pole is over the continent of Antarctica.

- The poles have no permanent human inhabitants.

- The South Pole is covered with a continental glacier.

- A troubling ozone hole forms over the South Pole every spring.

Part 5

A Global Overview

Now that you have completed your tour of the world's developed and developing regions, it's time to stand back and look at the big picture. Although you can easily get caught up in the lines that define the boundaries of regions and countries, remember that if you take a stroll on the physical landscape, those lines don't really exist.

Part 5 of this book looks at issues that don't fit neatly into a discussion of a single country or even a single region. These global issues affect us all. More important, it will take international cooperation to solve some of the problems we have created for ourselves. We all have to be informed and take responsibility for our planet—the ultimate geographic task!

"Looks like mostly water to me."

Chapter 27

Coping with Population Growth

In This Chapter

- ◆ Understanding population rates—people everywhere

- ◆ Finding where the people are—and where they're not

- ◆ Coming to grips with the food dilemma

- ◆ Taking a step in the right direction: demographic transition

In many of the regions explored in this book, population pressures are straining the environment. Although some regions have relatively small populations, their land is typically unable to support large numbers of people. In some of the most fertile areas of the earth, millions of people pack into the usable land, creating massive clusters that dot the landscape.

At one time in human history, it was possible for overcrowded populations simply to move elsewhere. Today, the earth is running out of places to put people.

As the human population continues to expand, every resource on Earth is taxed. Many environmentalists consider the world's population problem to be the greatest environmental threat—because the sheer volume of the world's people lies behind virtually every other environmental problem humans face. Although the earth has shown remarkable resiliency, are humans stressing even the earth's ability to heal itself?

A Little About Demographics

Geographers look closely at demographics, the study of population, because it has so great an influence on the workings of the earth. They're particularly concerned with the distribution and movement of populations.

The field of demographics is based largely on rates: birthrates and death rates are used to determine rates of population increase. The birthrate gauges the annual number of live births per thousand people; the death rates indicate the number of annual deaths per thousand people. To determine the natural growth rate, you simply subtract the death rate from the birthrate. The remainder is the increase (or, in rare instances, the decrease).

Population growth occurs when the birthrate exceeds the death rate. The greater the difference, the larger the growth. Population growth can occur because of an increase in the birthrate, a decrease in the death rate, or both. The only way to decrease population growth is to increase the death rate (not a people-friendly option) or decrease the birthrate.

Terra-Trivia

In the year 1 C.E., about 300 million people were living on the earth. This figure did not double to 600 million until the year 1650, which means that it took 1,650 years for the world's population to double. The population took only 200 years to double again, and then again in under 100 years, and most recently the doubling time was only 45 years.

Six Billion and Counting

In January 2007, the world's population reached 6.57 billion people. Some people might say that although 6 billion plus people is a large number of people, the earth is a large place and can handle that number. Although the earth *is* large, according to the Population Education Training Project, the earth doesn't really have that much space.

About three fourths of the earth's surface is water. About one half of the land surface in uninhabitable, such as the poles, mountains, swamps, and deserts. About three fourths of the habitable land (about one eighth of the total) can't be used to grow food because the land is too wet, dry, cold, steep, rocky, or just plain infertile. To top it all off, humankind has covered a lot of good land with buildings and concrete. All that remains to produce food is one twelfth of the earth's surface. The top five feet of topsoil on this one twelfth remnant represents the entire amount of the earth's food-producing land for more than six billion people—and the topsoil is eroding and washing out to sea at an alarming rate!

The world's population is expected to reach 7 billion by 2013, 8 billion by 2028, 9 billion by 2054, and then perhaps stabilize at 10 billion by 2100.

The Scattered Masses: Where the People Are

Geographers look at not only the numbers of people on Earth but also their distribution. People are not distributed evenly across the face of the earth. The earth's largest population clusters are in South Asia and East Asia. The population of Asia represents more than half the world's total population. Although other, smaller clusters exist around the world, vast areas of the earth are essentially uninhabitable.

Just as population distribution varies dramatically around the world, so too does food production. The ability to grow food can depend heavily on the length of the growing season, rainfall, soil fertility, and available technology. Droughts, floods, disease, war, and pest infestation can decimate food production in an entire region. Although international aid can do much to help in times of food crises, it's often too little and too late.

In this century, horrible food shortages have caused starvation and suffering at previously unheard of levels. Each day, hundreds of millions of the earth's people (perhaps one eighth of the world's population) go to bed hungry. The world's resources are being stretched, and desperate people are encroaching on marginal land. Despite the environmental consequences, rain forests are being cut and mountains are being terraced to create subsistence farming, causing serious erosion issues.

Although some countries, such as the United States and Canada, produce huge amounts of food for export, the food race also has its losers. Many countries in Africa, for example, already have more people than their land is able to feed. Even in a good year, these countries are dependent on imported food. Even more disturbing is that these same countries exhibit some

Terra-Trivia

Industrialized countries shouldn't blame the world's woes on the expanding populations of developing countries. Not so long ago, industrialized countries were the global leaders in population growth. Today, these countries gobble up much more than their share of the world's resources. On average, one person in the United States uses the same amount of energy as 400 people in the African nation of Angola.

of the world's highest growth rates. Perhaps more disturbing is that in 2006, for the sixth time in seven years, the world grew less food than it ate. In other words, we had to depend on dwindling food stocks to meet the demand for food. Although this practice of relying on foodstuffs stored during good harvests to be used during times of bad harvests has worked for untold centuries, the question now remains, "Will we have enough next year?"

More People, Less Food

Changing the population-growth trends in a region takes years (except in places with a state-mandated control policy, such as China). Because of the time it takes to decrease growth rates, especially in developing countries, humans soon will face increasing food shortages unless they're able to produce more food on Earth. Is this possible?

The problem can't be solved by bringing new land under cultivation. Most of the world's arable land is already being used, and the remaining land offers little potential for any sizable agricultural yields. With no more land to place under the plow, people must turn to methods of increasing production on lands already being used for farming, which is already being done in many areas. Although this section focuses on agriculture in its discussion of food resources, keep in mind that another major source is available: fish. More than 90 percent of all fish consumed on earth come from the oceans and seas. Not so long ago, most people viewed the immense oceans as an unending resource—but not anymore. Industrial fishing techniques have taken a heavy toll on the world's supply of fish. It has been estimated that since the beginning of large-scale fishing in the 1950s the populations of such large ocean species as cod, halibut, tuna, and swordfish have been reduced by as much as 90 percent. Although the growth of aquaculture, or fish farming, can make up for some of the loss, the days when the oceans could be regarded as a limitless resource for the world's hungry may be history.

The Green Revolution

Beginning in the 1950s, the world experienced a phenomenon known as the green revolution, a term that describes many scientific advances applied to agriculture to increase crop yields. Some technologies include bioengineered hybrid high-yield seeds, organic and chemical fertilizers, chemical pesticides, chemical herbicides, mechanization, high-tech irrigation systems, and multiple-cropping.

The application of green revolution techniques initially produced significant agricultural increases. Eventually, however, problems with the technology also became evident. One problem is expense: hybrid seeds, irrigation systems, machinery, and fertilizers and other chemicals are all expensive—and energy intensive. Although the developing countries stand to gain the most from adopting green revolution techniques, those countries have the least amount of money to spend on expensive technologies.

Another problem is that large portions of the population in developing countries typically are involved in small-scale subsistence farming. Small fields can't handle the same kind of mechanization used on the great plains of North America. Green revolution techniques produce the highest yields when they're applied to large expanses of a single crop.

The introduction in the mid-1990s of genetically engineered (GE) crops opened a new chapter in the age-old struggle to increase agricultural yields. New genetic material can now be added to crop plants to increase their resistance to disease and pests, control weeds, and even extend shelf life. But this scientific breakthrough has been controversial, with many people questioning the safety of GE crops and their impact on the environment, and many of the hungriest peoples reject the new foods due to cultural and taste considerations.

Disease, Rats, and Superbugs

One of the arguments in favor of genetic engineering is that it can increase a plant's disease- and pest-resistance and in some cases reduce the need for chemical pesticides. The opposite is true of hybrid seeds, which produce higher yields than indigenous crops, but are more vulnerable to local blights. Large areas of a single hybrid crop are much more susceptible to drought, pests, and disease than are small plantings containing multiple varieties of indigenous crops. To protect these high-yielding but vulnerable hybrids, farmers have turned to ever-increasing applications of chemical herbicides, pesticides, and fungicides.

In addition to the pollution and contamination these chemicals cause, they have only a short-term effect on the pest problem. The extent of the problem cannot be minimized: each year, pests account for the destruction of an estimated half of the world's food production. To increase food production, the loss to insects, rodents, and fungi must be minimized—but are chemicals the answer?

Indiscriminate chemical spraying kills not only the targeted pest species but also the helpful natural enemies of that species. Although spraying typically produces a short-term benefit, the pests eventually return in increased numbers. After natural controls are destroyed, pests can decimate crops with abandon. One possible alternative to the widespread use of chemical spraying is integrated pest management (IPM). This system addresses the entire ecosystem and is sustainable over long periods while remaining environmentally friendly. In addition to its selective and limited use of chemicals, IPM relies on the use of natural pest predators, low-till agriculture (minimal soil disruption), and crop diversity to control pests.

Eye on the Environment

A farmer using the IPM system accepts a small crop loss to pests in return for long-term sustainable yields. Many farmers would describe IPM as working with the ecosystem rather than against it. Although a farmer must be well educated about the subtleties of the local ecosystem, the lower costs of implementation are attractive to people in developing countries.

Although farmers can repeat the expensive spraying, of course, a higher dose is required the next time. Why? Because of the superbug phenomenon. Although most of the pests are killed during each spraying, some survive. The survivors have a natural immunity to the given level of toxin. These highly resistant survivors reproduce and breed a new generation of pests that are more resistant to spraying than the preceding generation. The only way to kill off these varmints is to hit them with yet a stronger dose of chemicals—and the cycle continues. The world eventually will become too toxic for human habitation—but will be just fine for the newest generation of superbugs.

A Glimmer of Hope: The Demographic Transition

Although population growth will continue to put pressure on the earth's resources in the short term, a phenomenon called the demographic transition could prove to be the answer for the long term, as shown in the following graph. To understand the demographic transition, you have to look at its three stages. The concept of demographic transition was developed around the population of industrialized Europe, which has passed through each stage.

Stage 1

Several hundred years ago, in preindustrial Europe, the population was relatively stable. The average woman in that society began having children early in life. If childbirth didn't kill her, she was likely to give birth to at least 8 to 10 children.

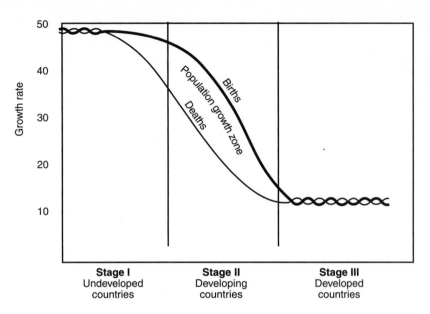

The demographic transition.

Birth control was unavailable and lots of children benefited an agricultural economy, as in developing countries today. In this type of society, children are an economic asset: they don't eat much, no education costs, and they can work in the fields and provide labor and income for their family. The typical eighteenth-century peasant farmer depended on his family to sustain him if he could no longer work the fields. If he had no family, his prospects for a happy retirement were dim. The conclusion: the bigger the family, the better.

Although you might think that the high birthrates of preindustrial Europe would give rise to high population growth rates, it wasn't necessarily so. Don't forget the death rates. Infant mortality was also extremely high. Perhaps half the children who were born never reached adulthood. People also died young by today's standards. Anyone who made it to age 40 was doing quite well. Poor hygiene, disease, famine, and war all took their toll. Death rates were high. With both high death rates and high birthrates, population growth was slow and fluctuated with war or famine.

Stage II

As the industrial revolution created new jobs and opportunities in Europe, the region entered demographic Stage II. New crops (such as potatoes and maize) were imported from the New World and proved to be a boon to agriculture. The situation was looking up as medical advances, improved nutrition, and hygiene increasingly enabled people to live longer lives. Although the death rate plunged, the birthrate declined more slowly. (Death rates can decrease dramatically within just a couple generations, but birthrates take much longer.)

Children still helped on the farm and went to work in the new factories. People still wanted to have large families, and infant mortality had dropped. In a family with 10 children, 8 might have made it to adulthood. The result of high birthrates, lower infant mortality, and lower death rates were huge population growth rates. Europe's population exploded.

Europeans were lucky. Although the population explosion had made the region somewhat crowded, people could always take off for the New World, Africa, and Australia. Europe weathered its population explosion through emigration. Regions lagging 100 or 200 years behind Europe in the demographic cycle would not be as lucky.

Stage III

Europe is now a highly developed region. Few people farm the land, and most live in cities. Death rates are even lower now, with people living well into their seventies. Social Security and birth control are available, and most people live in apartments, not on farms.

The net result: raising children is expensive, and Europeans have stopped having large families. The economic tables have turned, and a large family is no longer an economic windfall; it's an economic liability, in fact. In Europe, families are small: one or two children is the norm. The result of low death rates and low birthrates is stable populations. Some European countries are even experiencing declining rates—which hasn't happened since the black death. The decline is steepest in Russia and parts of Eastern Europe. The disintegration of the Soviet Union in 1991 and the upheavals that followed led to a spike in death rates and a plunge in birthrates in many of the former Soviet republics and satellite nations. The resulting negative growth rate persists in the region despite increasing economic and political stability.

Slowing Growth

The force that drives the demographic transition is economics. The more economically advanced a society becomes, the lower its growth rate. Stable, and even declining growth rates, are possible under the proper economic conditions. Can this model be applied to today's developing countries? There are signs that this is already happening.

World population is expected to continue to skyrocket throughout this century—but at a slower rate than in the last century. It will take increasingly longer periods to add an additional billion people. After peaking in the late 1960s at 2.04 percent, the world's population growth rate fell to 1.2 percent by 2002.

Thanks to the increased availability and use of contraceptives, couples in developing countries are now having an average of three children, down from six in the 1970s. Even China, once the world's demographic time bomb, has brought its growth rate down to 0.87 percent, although the methods used to reach this level were often coercive and unacceptably harsh. An even grimmer factor in the slowing of world population growth is the rising death rate from HIV/AIDS, especially in sub-Saharan Africa where the AIDS crisis is most severe.

Slowing growth rates by no means signal the end of the population problem. The vast number of women of childbearing age in developing countries means that populations there will continue to rise, in many cases unsustainably. At the same time, the world's population is growing older, international migration from developing to developed nations continues to rise, and more and more people are concentrating in vast megalopolises. These demographic trends will have wide-ranging economic, social, and environmental consequences throughout this century.

Controlling the growth of the world's population is critical to ensuring the future security and well-being of humankind. However the population dilemma is resolved, it won't be by just a handful of countries, but by all the nations of the earth acting together.

The Least You Need to Know

- Demographics is the study of population numbers, growth, distribution, and trends.

- More than 6.5 billion people live on the earth; by the end of the twenty-first century that number will likely have climbed to more than 10 billion.

- Demographic transition describes the lower growth rates that come with improved economic conditions in an urbanized society.

- The world's population will continue to rise throughout the twenty-first century, but at slower rates.

Saving Our Planet's Atmosphere

In This Chapter

- ◆ The atmosphere, in trouble?
- ◆ Keeping an eye on global warming
- ◆ Acid rain and nuclear accidents

Each region on Earth—land, sea, and sky—has its own environmental problems. Every nation, region, and city has unique concerns. The question is not identifying the problems; rather, it is determining what mankind will do to fix, or at least slow down, the problems.

Surrounding the earth is a thin layer of gases that's essential to all life. Humans unfortunately have not always treated that layer, the atmosphere, kindly. The atmosphere is one of those universalizing agents. Many atmospheric problems that trouble one part of the world will eventually trouble other areas, or even the whole planet. Because the air on which all humans depend is constantly mixing as it swirls around the earth, human problems are also distributed.

If people hope to solve the array of atmospheric ills they face, they must learn to overcome their differences and work together, because it is not a local or regional problem, but a universal one threatening our very existence.

Anatomy of the Atmosphere

The earth's atmosphere is made up mainly of nitrogen (78 percent) and oxygen (21 percent), as well as smaller amounts of other gases and water. It extends for nearly 350 miles from the surface of the earth (though it just gradually fades away and does not have a distinct boundary with outer space) and is divided into several distinct zones, or layers. The layer closest to the earth is the troposphere. "Tropos" is a Greek word meaning "turn," which is exactly what happens in this layer. The storms, temperature swings, winds, and other weather disturbances experienced on the earth's surface all take place in the troposphere. This zone encompasses the entire earth, although it's deeper at the equator and a little shallower at the poles. As you move higher in the troposphere, the temperature decreases.

About seven miles up, a thin transitional zone called the tropopause separates the troposphere from the next layer, the stratosphere, home of the famous ozone layer, which absorbs and deflects ultraviolet radiation from the sun (more about this later in this chapter). The stratosphere is drier and less dense than the troposphere.

The earth's atmospheric zones.

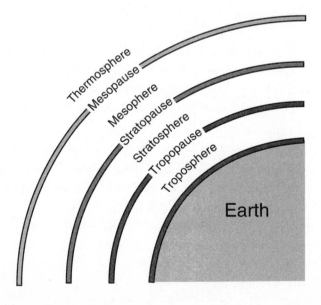

At about 30 miles up, the stratopause divides the stratosphere from the mesosphere. Within this layer, temperatures again decrease with height until, at 53 miles up, you reach another transitional zone, the mesopause, and beyond that, the thermosphere, or upper atmosphere. This is the deepest layer of the atmosphere, reaching some 372 miles above the earth's surface. Temperatures here remain constant for a while and then begin to increase with height due to the effects of solar energy.

Turning Up the Heat: Global Warming

Global warming is one of today's most intensely discussed environmental topics. Signs that the earth's climate is warming are there for all to see (and feel), and they are mounting. Global surface temperatures increased 1.1 degrees F over the last century, and 0.4 degrees F in just the last 25 years. Mountain glaciers are receding, polar icecaps are thinning, and sea levels rose at a faster rate during the twentieth century than in the preceding several thousand years. The question is, what's causing it all? On one side of the debate are those who believe that this warm spell is just part of a natural and age-old cycle of global climate change. On the other side, the majority of climatologists point to human activity, specifically the burning of fossil fuels, as the primary culprit. If human activity is indeed to blame for global warming and nothing is done to reverse the trend, the results could be catastrophic to human life.

The Greenhouse Effect

Despite all its bad press, the process known as the greenhouse effect is a very natural and necessary phenomenon. To understand it, think about the way a greenhouse works. Light from the sun passes through the glass of a greenhouse and is absorbed by the earth and plants within. The ground surface reradiates the energy as infrared radiation, or heat energy.

Although the short-waved sunlight initially passed through the greenhouse glass, the reradiated long-wave infrared heat energy is not able to pass back out efficiently. This heat energy becomes trapped and causes the greenhouse to warm. (The same process causes your car to get unbearably hot while sitting in the mall parking lot on a summer day.) The earth's atmosphere acts in much the same manner as a greenhouse.

The atmosphere contains gases that enable sunlight to pass through to the earth's surface but hinder the escape of the reradiated heat energy. Without this natural greenhouse effect, the earth would be much too cold to sustain life. The problem arises when too much of the heat-trapping gases are present in the atmosphere. Then temperatures begin to rise, and the troubles begin.

What Causes the Greenhouse Effect?

Naturally occurring gases in the atmosphere—carbon dioxide, methane, nitrous oxides, and water vapor—account for most of the greenhouse effect. The most abundant of these gases is water vapor, followed by carbon dioxide. Human activities such as agriculture, industry, mining, and especially the burning of fossil fuels, have increased the levels of some of these gases, and added others that are not naturally occurring, such as chlorofluorocarbons (CFCs). Although CFCs and methane are more efficient than carbon dioxide at trapping energy in the atmosphere, carbon dioxide is a greater cause for concern for two reasons: it is far more abundant in the atmosphere, and humans add so much more of it through the burning of fossil fuels. In fact, carbon dioxide is thought to be responsible for more than 50 percent of human-induced warming. All life is based on carbon. The coal, oil, and natural gas that humans rely on so heavily to fuel their way of life were once the living carbon-based tissues of primeval plants and animals. Buried for eons under layers of sediment, these ancient organic remains were eventually transformed by heat and pressure into the fossil fuels people use today. When those fuels are burned, they release the carbon that had gone into their making back into the atmosphere.

Carbon is also released when a forest catches fire, when you burn charcoal in a grill, or even when you take a breath. Humans survive by breathing in oxygen and releasing carbon dioxide. Plants do the exact opposite, taking in carbon dioxide and releasing oxygen through the process of photosynthesis. This mutually beneficial exchange between plants and humans has the added advantage of helping to control carbon dioxide buildup in the atmosphere. But plants can do only so much to absorb all the carbon dioxide generated by the burning fossil fuels. And deforestation and associated forest burning is doubly harmful in terms of exacerbating the greenhouse effect (in 1997, Indonesian fires put 13 to 40 percent as much carbon dioxide into the air as did fossil fuel burning in that year).

Consequences of Global Warming

Projections based on computer climate modeling raise disturbing possibilities of global warming. The three most probable include …

- ◆ **Rising sea levels.** In fact, this is already happening. A further increase in global temperature could cause the massive polar ice caps (especially in Antarctica) to melt. The huge influx of water from the ice caps would cause ocean levels to rise even faster than they are now. Ocean temperatures would increase, causing thermal expansion of the water molecules and an additional rise in sea level.

Some of the world's largest population clusters exist along coastal areas, and rising sea levels would obliterate them.

◆ **Changing climates.** Most people could survive if their hometown became a little hotter or drier. Unfortunately, agriculture around your town might be altered dramatically. Because temperature and precipitation would not be uniformly affected by global warming, the outcome of this scenario is difficult to predict. But if the Great Plains of the United States became too hot and dry to grow grain, not only would residents be in trouble, the impact on the world's food supply would be severe. Considering the earth's burgeoning population, climatic threats to the food supply are worth worrying about. And the problem isn't limited to farmlands. Some of the earth's most productive and fragile ecosystems—corals reefs, mangrove swamps, and tropical rainforests—are especially vulnerable to climate change.

◆ **Extreme weather.** Ten of the hottest years on record have occurred since 1987. The hottest year so far was 2006, which corresponded with the strongest El Niño on record and a host of weather related disasters: devastating floods, prolonged droughts, powerful hurricanes, and tropical forest fires. More persistent El Niños, more severe droughts and floods, and an increasing incidence of devastating weather events are all projected consequences of unchecked global warming.

Is Global Warming Fact or Fiction?

While most scientists agree that the earth is warming, some scientists assert that increases in greenhouse gases will eventually result not in a global temperature rise but rather in an icebox effect, in which global temperatures will drop.

According to this theory, as the earth warms, ocean water evaporates more quickly, resulting in an increase in cloud formation. If the clouds form as a low, dense layer, the more extensive cloud cover will act to screen and reflect incoming sunlight, ultimately causing surface temperatures to drop. This is a classic example of negative feedback, in which a mechanism that alters the earth's climate equilibrium in one direction (for example, increased warming) generates a response in the opposite direction.

Although some scientists debate whether there is a human-induced global warming crisis, a consensus is building among the nations of the world that something needs to be done. The Kyoto Treaty, drawn up in Kyoto, Japan, in 1997, committed industrialized nations to reducing their overall emissions of greenhouse gases by an average of 5.2 percent below 1990 levels within 10 years—a far from adequate response according to some environmentalists, but a start nonetheless. As of December 2006, the treaty had been signed and ratified by 166 nations, including the European Union, Canada, and Japan. Only the United States, the world's biggest producer of greenhouse gases, and Australia have signed but declined to ratify the treaty—in fact, withdrawing from the treaty in 2001, citing its negative economic impact.

Although governments may not agree on a cure, it is clearly in our interest to limit our consumption of fossil fuel, use clean-energy technology whenever possible, and encourage the conservation of our forests. These practices are sound, no matter what the outcome of supercomputer global-warming models might be. The conservation of fossil fuels will ensure an energy supply well into the future. Saving forests not only cleans the air but also helps to preserve biodiversity on the planet.

Ozone Depletion, or Pass the Sunblock!

Before looking closely at the atmospheric problem of ozone depletion, you should be able to distinguish between the ozone layer and ozone smog. In the lower layer of the earth's atmosphere (the troposphere), ozone is a pollutant. A by-product of auto exhaust and sunlight, ozone builds up to produce ozone smog that often menaces cities during the sunny and warm summer months. Although this form of ozone is a health threat, don't confuse it with the ozone in the upper atmosphere.

The ozone in the stratosphere is essential to life on this planet because it protects humans from the harmful effects of the sun's ultraviolet radiation. "Ozone depletion" refers to the degradation of this protective shield in the stratosphere. Unfortunately, humankind has devised several chemicals that, when released to the atmosphere, attack and degrade the protective ozone shield. The most notorious chemical family, responsible for the bulk of the world's ozone destruction, is the chlorofluorocarbons, or CFCs.

CFCs are found in foam packaging, refrigerator and air-conditioner coolants, and insulation and are used as an industrial cleaner. When CFCs are released, they break down the molecules of ozone in the stratosphere at an alarming rate. The reduction over the South Pole each spring is so severe, in fact, that it's called the "ozone hole." Without an ozone layer or with a reduced ozone layer, much more of the sun's ultraviolet radiation reaches the earth's surface. Ultraviolet radiation can be extremely harmful to life on Earth in a number of ways. An excess of this radiation can damage plant life, which produces oxygen for humans. This radiation is also worrisome in its effect on the minute phytoplankton that abounds in the earth's oceans. Phytoplankton, which is at the base of the ocean's food chain, is a tremendous consumer of carbon dioxide and producer of oxygen.

The direct effect on humans has also become increasingly apparent in recent years. Increases in skin cancer, eye cataracts, and damage to the human immune system are just a few of the consequences. Unlike the global warming issue, there is general scientific consensus on the link between CFCs and ozone depletion. The 1987 Montreal Protocol, which originally committed signatory nations to reducing their use of CFCs, was later amended to ban CFC production after 1995 in developed nations and by 2000 elsewhere. Unfortunately, these long-lived chemicals continue to deplete ozone long after their initial release to the atmosphere. Even with a total ban on CFC production, the ozone layer is not expected to fully recover until about 2050.

That Burning Sensation: Acid Rain

Often referred to as "death from the sky," acid rain is a broad term used to describe the various ways acid falls out of the atmosphere. In addition to acid rain, there's also acid snow and acid fog, and about half of the acidity in the atmosphere falls back to earth not as moisture but as particles or gases that are "dry deposited" on buildings, cars, and trees. In the past several decades, acid rain has become a persistent and global problem. Its primary causes are industrial sulfur dioxide emissions and nitrogen dioxide from automobile and truck exhausts. Placed into the atmosphere and mixed with water vapor, oxygen, and other chemicals, sulfur dioxide forms sulfuric acid, and nitrogen dioxide forms nitric acid.

The acidity that falls from the sky can result in a host of environmental problems. The increased acidity "eats" away at buildings, machinery, and living things.

When acidity levels rise in lakes and rivers, fish and aquatic life can no longer survive. The results are biologically dead lakes that are now prevalent in the Northeastern United States and in Eastern Canada. Acid rain, perhaps combined with other stresses such as insects, drought, disease, and other pollutants, is at least partially responsible for forest die-offs that are occurring worldwide. The Eastern United States and Western and Northern Europe have been hit especially hard.

The solution, of course, is to reduce the emission of sulfur and nitrogen dioxides. This reduction is not easily accomplished, however. Because weather systems often move the problem far from its original source, identifying a specific polluter is difficult. It's also difficult to legislate mandates that would require one country to invest billions of dollars in a cleanup when the impact of its pollution occurs in its neighbor's backyard. In the meantime, Canadians are none too happy about their country's thousands of dead lakes that have resulted largely from pollution originating in the United States.

Nuclear Fallout

Nuclear energy is not generally regarded as an atmospheric problem. However, that idea changed on April 26, 1986, when the number-four reactor at the Chernobyl nuclear power plant blew up. The roof of the containment building was blown off, and a deadly nuclear cloud rose into the atmosphere. The Soviets evacuated everyone within a 40-mile-diameter circle later known as the "zone of estrangement." The evacuation ultimately displaced about 116,000 residents of the area.

Although the accident occurred in what was then the Soviet republic of the Ukraine, the bulk of the radioactive fallout fell on nearby Belarus. As it circulated in the atmosphere, the radiation did not stop at national borders. Much of Western and Northern Europe received large doses of radiation from the accident. Even in the far Scandinavian north, the Lapp's reindeer milk was contaminated by radiation.

To help contain the contamination approximately 180 tons of uranium fuel and another 10 tons of radioactive dust remain sealed within a 24-story concrete-and-steel enclosure erected around the ruined reactor. The core of the reactor is still active, and the enclosure has so deteriorated that it is in danger of collapsing. Equally disturbing is the fact that two of the four original Chernobyl reactors continued to operate until 2000 and that 13 other reactors of the same flawed design are still on line.

The Least You Need to Know

- The health of the atmosphere affects not only individual nations or regions but the entire earth and all its inhabitants.

- The lowest layer of the atmosphere is called the troposphere. This is where humans live and weather happens.

- Global warming, resulting from the greenhouse effect, might already be causing global climatic change and rising sea levels.

- Ozone depletion is caused primarily by chemicals called CFCs. Loss of the ozone shield poses a health threat to all life.

- Acid rain is caused by industrial and vehicular emissions; dying forests and dead lakes are the result.

- The Chernobyl power plant disaster opened people's eyes to the potential global consequences of a nuclear accident.

Chapter 29

Meeting Energy Needs and the Water Challenge

In This Chapter

- ◆ The energy dilemma and costs of fossil fuels
- ◆ Why renewable resources matter, and what about hydrogen?
- ◆ The coming fresh water crisis

They say money makes the world go round, but what actually does the trick is energy. People need energy to run factories, drive cars, heat buildings, and operate just about every gizmo in the house. As society has become more industrialized, human energy needs have grown. The energy demands of the developed world far exceed those of developing countries, although the gap is narrowing. The United States is the most demanding country of all: with less than 5 percent of the world's total population, the United States consumes nearly 25 percent of the world's annual energy production.

World energy demand is expected to increase as much as 58 percent by 2025, with developing countries accounting for much of that increase. How that demand is met and at what cost to world peace and the health of the environment are key questions confronting humankind in the twenty-first century.

Nonrenewable Resources: Here Today, Gone Tomorrow

Most of the world's energy needs come from *nonrenewable* sources, mainly fossil fuels: oil (petroleum), natural gas, and coal. Nuclear energy, derived from uranium, is not a fossil fuel, but it is nonrenewable. With all nonrenewable energy sources, the bottom line is this: once you use it, it's gone for good. You can't replant it or regrow it, and supplies are not limitless.

def•i•ni•tion

Nonrenewable resources, such as fossil fuels, metal ores, and uranium used for generating nuclear power, take so long to form that they're essentially irreplaceable. Although wood is generally considered a renewable resource, if a cleared forest is not replanted, its wood automatically becomes a nonrenewable resource.

The Old Fossils

The world today is largely powered by fossil fuels, which account for 85 percent of total energy consumption. The benefits of fossil fuels are easy to understand. They are still available in vast quantities and they are relatively cheap to extract from the earth, but although supplies might be abundant today, they won't always be. No one can say for sure when the world will literally run of gas, but that day will definitely come—and sooner rather than later if current consumption trends are not reversed. The energy shortages of the 1970s scared many people into conserving fuel, but the effects were short-lived, as oil prices dropped again in the mid-1980s and gas-guzzling cars came back into fashion.

Although extracting fossil fuels might be relatively cheap, there are many hidden and not-so-hidden costs involved in the process. Mining, drilling for, and transporting fossil fuels can all have devastating effects on the environment. Strip mining and oil spills are only two of the more visible examples. Once fossil fuels are out of the ground and in your gas tank or local power plant, they are adding greenhouse gases and other pollutants to the atmosphere.

Finally, there is the political side of the fossil fuel equation. The world is addicted to oil, and much of it is sitting under one of the world's most dangerous powder kegs. More than half of the world's estimated oil reserves are in the Middle East, mostly in the Arabian Peninsula, Iraq, and Iran. The Arab oil embargo of 1973, the Gulf Wars against Iraq of 1991 and 2003, and the escalating political instability and terrorist violence in the region are all clear indicators of the need to shake our global oil addiction and find alternative fuel solutions.

The Nuclear Card?

The newest of the nonrenewable energy sources, nuclear power plants, use the energy that is released by the splitting of uranium atoms to generate electricity. The first nuclear power plants went on line in the 1950s. Today, nuclear power accounts for about 16 percent of the world's electricity (20 percent in the United States). Once hailed as a safe, clean alternative to fossil fuels—the energy of the future—nuclear power is in decline. Few new plants are being built, and many are being phased out. The accidents at Three Mile Island in the United States and at Chernobyl in what was then the Soviet Union opened the eyes of many to the dangers of nuclear power plants. In addition to the threat of accidents, there's also the sticky problem of disposing of spent fuel and low-level radioactive waste. And in the aftermath of the September 11, 2001, terrorist attacks against the United States, nuclear power plants and nuclear waste storage sites are being viewed with mounting concern as vulnerable terrorist targets.

Renewable Resources: Here Today, Here Again Tomorrow

Renewable resources continuously regenerate or can be recouped within a relatively short period of time. Tree stands that are harvested and replanted can regenerate within a lifetime. In addition to wood and other types of organic biomass, the main renewable energy sources are flowing water, wind, sunlight, and geothermal energy.

Hydroelectric Power—Dam It!

Hydroelectric power, which uses the force of falling water to turn turbines and generate electricity, now accounts for about 20 percent of the world's electricity. Although hydropower would seem to be a safe, clean, and long-lasting solution to the world's energy problems, it presents its own problems.

First, potential sites for hydroelectric dams are limited and many have already been used. Second, huge reservoirs that build up behind hydroelectric dams can displace people and cause unforeseen ecological damage, as was the case in the Hydro-Quebec projects in Northern Canada, which displaced the Cree people and contaminated one of their primary food sources with mercury poisoning. China's controversial Three Gorges Dam, the world's largest hydroelectric power project scheduled for completion in 2009, is expected to force between 1 and 2 million people out of their homes.

Rivers typically flood during their annual cycle, cleansing the land within their floodplain and revitalizing it with nutrient-rich silt. Damming a river prevents the annual flooding, with long-term consequences for floodplain fertility. Silt that no longer reaches farmlands downriver collects behind the dam, often clogging turbines and affecting future power generation. These problems have been evident on such large-scale projects as the Aswan High Dam on the Nile River in Egypt.

Dams also have a negative impact on wildlife habitats and migration patterns. In the Western United States, for example, where most river systems have been extensively dammed, dams either have destroyed important salmon spawning habitats or keep the fish from reaching the spawning grounds that remain.

The Solar Solution

Although the earth has abundant, and free, energy bombarding daily from the sun, it is not cheap to convert it into usable energy consumption.

Energy from the sun is used in two principal ways: to generate electrical energy and to supply heat. Solar energy can be converted directly into electricity by means of photovoltaic cells. Used on rooftops and pocket calculators, they convert light to electrical energy, but are not cost efficient for larger projects (except in remote areas). As new techniques for manufacturing these photovoltaic cells are developed and as fossil fuel prices rise, the cost might soon be competitive with traditional power sources. These systems are becoming sophisticated enough to work well with even ambient light and will therefore be adaptable to most places on Earth (except maybe at the North and South poles).

Solar energy is also widely used for heating buildings. Many passive solar-heating techniques, such as orienting buildings to take maximum advantage of the sun's warming rays, are centuries old. Such traditional building principles have been adapted and combined with new techniques and materials to create energy-efficient designs at a minimal cost. In addition to passive solar techniques, active solar systems that collect and store solar power can also be used to heat buildings.

These active systems typically employ solar panels that absorb sunlight and heat fluids circulated through panels in pipes. The heated fluid is then used to warm living spaces or, more frequently, to heat water for the domestic hot-water supply. Solar panels work best, obviously, in areas that annually receive sufficient days of sunshine. Areas that are especially rainy or foggy, such as the Pacific Northwest in the United States or the British Isles, are worse candidates than drier areas.

Active solar systems also have industrial applications. Solar thermal concentrating systems work like super solar panels, using lenses and mirrors to concentrate solar radiation and produce extremely high temperatures that can be used to create steam and generate electricity.

Harnessing the Wind

People have used wind power for centuries to run mechanical equipment and pumps. The most notable examples are the beautiful old windmills of the Netherlands (Holland), which constantly pumped water up and out of canals used to drain land reclaimed from the sea (polders). Windmills (of higher-tech design than the old Dutch versions) are now used to turn generators for the production of electricity. (Because of more constant and higher-velocity winds found higher in the atmosphere, modern wind towers have gotten increasingly high. A 430-foot wind turbine was built in the United Kingdom.)

Some places are windier than others, of course. Windmills work well in the Netherlands because of the steady wind from the North Sea. Many places are simply too calm to make windmills a feasible option. They also tend to be noisy, and can add to visual "horizon pollution." Although it still supplies only a small part of the world's electricity, wind power is the fastest growing source of electrical energy, and it's becoming increasingly cost competitive. The world's leading producer of wind-generated electricity is Germany, followed by Spain. Growth in the United States has slowed because of the uncertain status of federal wind-power tax incentives.

Geothermal Energy and Biomass

Geothermal heat from deep within the earth can be tapped to create steam that turns turbines and generates electrical power. Geothermal energy is an important source of power in Iceland and nearly every home there is heated with water from hot springs.

Unfortunately, geothermal energy is not easily accessible everywhere on Earth. The largest geothermal facilities are located at geologic hot spots on the earth's surface, generally seismic areas, such as Iceland and along the Pacific Rim "Ring of Fire." The United States is the leading producer of electricity from geothermal reservoirs, with most of its plants concentrated in California and Nevada.

Another source of geothermal energy would more accurately be called ground source heat pumps. GSHP systems tap the constant ground temperatures that exist beneath the frost line to heat, cool, and generate hot water for commercial buildings and homes. These systems run on electricity and can be remarkably efficient depending on latitude, soil, and rock type at a given site.

Another major renewable source of energy worldwide is biomass, which includes wood and other plant materials, organic garbage, and animal wastes. Biomass was the first source of energy used by humans and it remains a primary fuel in many developing countries. The most direct way to extract energy from biomass is to burn it for cooking, heating, or industrial purposes, but that releases carbon dioxide into the atmosphere, just like burning any other carbon-based fuel would (except that these sources are not releasing prehistoric carbon, but only releasing the carbon the plant has absorbed during its lifetime; this makes biomass carbon neutral), and it's inefficient. New technologies, however, are making it possible to use biomass in far more efficient and environmentally friendly ways. Biomass can be converted, for example, into cleaner burning combustible gases and liquid fuels, such as ethanol, which is being used to help reduce gasoline consumption. Two promising scenarios for the future include the large-scale farming of fast-growing grasses and trees as biomass "energy crops" and the use of anaerobic digestion to generate methane gas.

The Dawning of the Hydrogen Economy

It might still be only a dream, but a lot of scientists are dreaming it. What's the dream? A world fueled by a limitless energy source, powerful, efficient, and clean burning, whose most noxious emission is pure water. Sounds like a fantasy, right? Well it isn't that far-fetched. The magic fuel is hydrogen, the lightest, simplest, most abundant element in the universe. And right now, plying the streets of smog-filled cities around the world, are experimental fleets of buses running, pollution-free, on hydrogen fuel cells.

Hydrogen, Hydrogen Everywhere

Literally, hydrogen is everywhere. Although it doesn't exist naturally on Earth as a gas, wherever there's water (H_2O), there's hydrogen (H_2). It is also a component of natural gas and can be extracted from biomass. So why isn't the world running on hydrogen? The main hitch, for now, is the cost of separating hydrogen from the compounds in which it is found. The technology for splitting ordinary water into hydrogen and oxygen is well-known, but it requires a lot of electricity and is costly.

Most of the free hydrogen used today is created by a process called "steam reforming," in which high-temperature steam is used to separate natural gas into hydrogen and carbon. But reforming is energy intensive and reliant on fossil fuels. The challenge for all those scientists dreaming of a hydrogen future is to devise a cost-effective, environment-friendly way of liberating hydrogen.

Drinking from the Exhaust Pipe: Fuel Cells

Hydrogen can be used like gasoline to fuel a combustion engine, but it really comes into its own when it's used with a device called a fuel cell. Put very simply, a fuel cell is a sort of battery in which a catalyst causes hydrogen molecules to separate into protons and electrons. The electrons flow out as electric current, the protons combine with oxygen molecules from the air to form pure water, the only residue of the process. In addition to powering vehicles, fuel cells can also be used to generate electricity and heat.

Although fuel cell technology is not new, the industry is still in its infancy and costs are high. But as the technology advances and hydrogen consciousness grows, costs are bound to come down to competitive levels. In a world faced with so many energy challenges, the potential of hydrogen is simply too great to ignore.

Taxing the World's Oceans

All life on Earth sprang from the oceans. Because oceans cover 70 percent of the entire earth, it's difficult to imagine that something so huge could ever be threatened. But they are.

The world's creeks, streams, and rivers all comprise a huge highway to the sea. When you fertilize your lawn, for example, rain washes the chemicals into storm drains. From there, they flow into streams and then rivers, and eventually out to the sea.

And so it is with sewage, industrial waste, and leaching landfills. Sometimes no one bothers with the indirect path: they just take it out to sea and dump it directly.

People used to think that the oceans could never be mucked up because they were simply too big. Many countries and municipalities simply cart their wastes out to sea and dump them. Although the type of waste varies, it might include treated or untreated sewage, industrial waste, garbage, construction materials, and even medical waste. Although most people want to forget about this refuse as it collects in the world's oceans, the stuff has been coming back to haunt us.

Terra-Trivia

Most oil pollution in the oceans is caused by conscious acts, mainly pouring used engine oil down drains, oily road runoff, the routine flushing of ship bilges, ballasts, and tanks, and natural seepage. Only about 5 percent comes from large tanker accidents and big spills.

As bad as these wastes are, they can seem almost benign next to the really hideous materials humans spew into the oceans, such as toxic chemicals, radioactive waste, heavy metals, oil, and PCBs (polychlorinated biphenyls, which have been linked to birth defects, cancers, and immune system damage).

Although solutions are possible, they require a will and a wallet. Only 19 years ago, Boston Harbor was the cesspool of the East Coast. Sewage from Boston and its surrounding communities was dumped directly into the harbor with little or no treatment. Political clout was mustered, a budget was hammered out, and billions of dollars and years of labor later, a huge new sewage-treatment facility stands on Deer Island at the mouth of the harbor. It's now possible again to find porpoises swimming in the clear blue waters of Boston Harbor.

Water, Water Everywhere and Not a Drop to Drink

Except for air, no other resource is as crucial to life on Earth as fresh water. A reliable supply of clean, fresh water is essential for drinking, cooking, and sanitation, but humans also rely on fresh water for transportation, agriculture, industry, and food (more than 40 percent of the earth's fish species live in fresh water). On a planet whose surface is mostly water, you might think that water scarcity would be the least of our problems. Not so. Only about 2.5 percent of the earth's water is fresh. More than two thirds of that is frozen in glaciers and permanent snow cover, while the rest—the accessible water in lakes, rivers, and aquifers (ground water)—is very unevenly distributed. Some regions enjoy abundant supplies; others are water-stressed.

Dwindling Supply + Growing Demand = Crisis

Freshwater supplies are not only relatively restricted to start off with, they are becoming scarcer. Rising global temperatures, more frequent and intense droughts, and changing rainfall patterns are straining water resources in many parts of the world. An even bigger culprit is pollution. Today, an estimated 1.1 billion people drink water that is unsafe.

As water supplies decrease, population growth and industrial and agricultural expansion are rapidly increasing worldwide water demand. Agriculture is a particularly insatiable consumer of water, accounting for about two thirds of the world's fresh water use. It is also highly inefficient. About 60 percent of the water used for irrigation is lost to evaporation before it ever gets into a plant. As with other resources, the United States leads the world in water use, to the tune of about 620 quarts per person per day. (The average African, by contrast, uses barely 50 quarts a day.)

Quenching the World's Thirst

Although seawater desalination (the process of removing the salt from salt water to make it potable) offers some hope for the long term, the worst-case scenarios could easily come true in the short term, unless the nations of the world get together to rethink water management policies, promote more efficient farm irrigation techniques, and curb the pollution that is poisoning the planet's lifeline.

Disappearing Forests: A Global Concern

Deforestation is a major concern. Not only the world's tropical rain forests, but temperate forests have also been under the ax. Although the global rate of deforestation slowed in the late 1990s, current estimates suggest that an area of forest the size of Maine is cut each year around the world.

Eye on the Environment
In the Asian mountain country of Nepal, forests are being cut at an alarming rate. Extreme poverty has forced people to clear steep hillsides in search of marginal farmland and firewood. The result has been a devastating loss of topsoil. Without tree cover to control runoff, flash flooding has increased, and Bangladesh, downstream from Nepal's rivers, has been subjected to yet another source of flooding.

The world's largest remaining concentrations of tropical rain forests exist primarily in three countries: Brazil, the Democratic Republic of the Congo, and Indonesia. These forests are being rapidly exploited, and the poverty in those countries contributes to slash and burn agriculture, making the problem more acute.

The impact of deforestation is far-reaching:

◆ **Atmospheric carbon dioxide buildup:** Trees remove carbon dioxide from the atmosphere and add oxygen. When you remove trees, you also remove a major atmospheric cleanser. The burning of trees releases carbon dioxide into the atmosphere, adding to the greenhouse gases that are responsible for global warming.

◆ **Topsoil loss:** Trees hold precious topsoil in place. Without forests, the land is subject to erosion and soil loss. The quick release of runoff rainwater is responsible for flooding.

◆ **Climate change:** Forests play a vital role in the earth's hydrographic (water) cycle by taking in rainwater (absorption) and giving off water vapor (transpiration). The cutting of trees can result in altered patterns of precipitation.

◆ **Species decline:** Forests provide the habitat for an unrivaled diversity of plant and animal species. Cutting even small parcels of forest can result in species extinction. A loss of species results in diminished biological diversity on Earth and a loss of potential sources for future foods and medicines.

The aesthetic component of tree loss cannot be minimized. For most people, a world without forests would be a much poorer place.

What can be done? Even if governments aren't doing much, you can ease the problem. You can support the many rain forest industries that help to make the forests economically viable without resorting to cutting. You can avoid using tropical hardwoods or woods (such as redwood) from old-growth forests. You can reuse or recycle wood-based products. You can plant a tree.

The Vegetarian Variation

Vegetarian eating habits are part of an environmentally friendly lifestyle.

A person's diet can affect the environment in many ways including changing how land is used. For instance, a fertile plot of land can support 20 times as many people on a vegetarian diet as on a typical American diet. Similarly, fewer tropical forests in

Central America would be cut down to support cattle grazing if people reduced the amount of beef they eat. And because more than half of all water consumed in the United States is used in the livestock industry, we could also help with the clean water issue by simply eating less animal protein.

Putting Your Money Where Your Mouth Is

Money not only talks, it influences behavior. One way to curb environmental ills is to make them costly for the perpetrators through the use of the tax code. In a movement that is gaining ground in Europe and elsewhere in the world, governments are experimenting with "environmental tax shifting," which penalizes pollution and resource waste while reducing the tax burden on workers and investors. The result can be a win-win for society: economic stimulus and a cleaner environment. In the United States, "green" taxes have been successful in phasing out ozone-destroying CFCs.

The Least You Need to Know

- Nonrenewable energy sources are finite in quantity, and their use presents pollution and disposal problems.

- Most of the world's oil reserves are in the volatile Middle East, where they are vulnerable to war, political upheaval, and terrorist threats.

- Renewable energy sources—water, sunlight, wind, geothermal heat, and biomass—offer a long-term alternative to fossil fuels.

- Hydrogen is an abundant, efficient, and clean-burning fuel with the potential to help the world finally break its fossil fuel addiction.

- The world's oceans are being contaminated at an unprecedented rate and the world is experiencing a scarcity of fresh water.

- Forest cutting has global consequences that will take a universal effort to abate.

Index

Q-R

S

BOOST YOUR BRAIN POWER

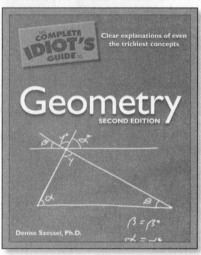
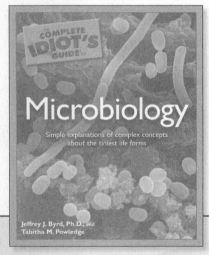